Concise Catalog of 1

Springer

London
Berlin
Heidelberg
New York
Hong Kong
Milan
Paris
Tokyo

W.H. Finlay

Concise Catalog of Deep-sky Objects

Astrophysical Information for 500 Galaxies, Clusters and Nebulae

With 18 Figures

 Springer

Cover illustrations: Background: NGC 2043, by courtesy of Zsolt Frei, from CD-ROM Atlas of Nearby Galaxies, copyright © by Princeton University Press, reprinted by permission of Princeton University Press. Inset 1: NGC 3031, by courtesy of Zsolt Frei, from CD-ROM Atlas of Nearby Galaxies, copyright © by Princeton University Press, reprinted by permission of Princeton University Press. Inset 2: M80, courtesy STScI. Inset 3: NGC 2244, by courtesy of Travis Rector and the NOAO/AURA/NSF. Inset 4: NGC 6543, courtesy STScI.

British Library Cataloguing in Publication Data
Finlay, W.H.
   Concise catalog of deep-sky objects : astrophysical information for 500 galaxies, clusters and nebulae
   1. Galaxies – Catalogs 2. Galaxies – Clusters – Catalogs 3. Stars – Clusters – Catalogs 4. Nebulae – Catalogs I. Title
   523.8'0216
ISBN 1852336919

Library of Congress Cataloging-in-Publication Data
Finlay, W.H. 1961–
   Concise catalog of deep-sky objects : astrophysical information for 500 galaxies, clusters and nebulae / W.H. Finlay
      p.cm
   Includes index.
   ISBN 1-85233-691-9 (acid-free paper)
    1. Galaxies – Observers' manuals 2. Galaxies – Clusters – Observers' manuals 3. Nebulae – Observers' manuals 4. Astronomy – Observers' manuals I. Title
   QB856.F56 2003
   522–dc21   2002036632

ISBN 1-85233-691-9 Springer-Verlag London Berlin Heidelberg
a member of BertelsmannSpringer Science+Business Media GmbH
http://www.springer.co.uk

Typeset by Florence Production, Stoodleigh, Devon
58/3830-543210   Printed on acid-free paper   SPIN 10891233

Acknowledgements

This book would not have been possible without the help of many individuals. Thanks are due to Alister Ling and Doug Hube who read and suggested changes to drafts of the book. Thanks are also due to Chris, Paul and Jenise Finlay for their help in removing as many typographical errors as possible. Thanks also to the many fellow amateur astronomers over the years, too numerous to list here, from whom I have learned so much. Finally, I thank my parents for instilling a boundless curiosity in me, and my wife and children for the kind patience and support they showed while I wrote this book.

Contents

1 Introduction 1

2 The Messier Objects 7

3 NGC (New General Catalogue) Objects. 55
   NGC 40–936. 55
   NGC 1022–1999. 70
   NGC 2022–2985. 83
   NGC 3003–3998 114
   NGC 4026–4995 141
   NGC 5005–5982 180
   NGC 6093–6994 191
   NGC 7000–7814 228

4 IC (Index Catalogue) Objects. 243

Index. 245

Introduction

This book is intended to give a concise summary of some of the more interesting astrophysical facts that are known about objects commonly observed by amateur astronomers. Pondering this information while viewing an object in the field has added a new level to the author's enjoyment of deep-sky observing, and it is hoped this information will be similarly enjoyed by other amateur astronomers. The book is not intended to be read cover to cover, but rather is designed so that each object entry can be read individually one at a time and in no particular order, perhaps while at the eyepiece.

A total of 520 deep-sky objects are listed as separate entries in this book, in order of their NGC (New General Catalogue) number in the main section of the book, including all the Messier objects, the Herschel 400 objects and the Royal Astronomical Society of Canada's Finest 110 NGC objects. Because NGC numbers were originally assigned in approximate order of an object's location from west to east, objects that are well placed for viewing in the sky at a particular time of year all occur within a few pages of each other.

For convenience, the Messier objects are repeated in a separate chapter in order of their Messier number. The only two objects from the IC (Index Catalogue) are listed in the last chapter of the book.

The following notes apply to the presentation of information for each object.

Object type: This is one of the following:

Open cluster – a close-knit collection of stars within the disk of our Galaxy that all formed from the same interstellar cloud in the past few billion years, containing fewer (usually many fewer) than a few thousand stars and often containing only a few tens of stars that are visible in an amateur telescope.

Globular cluster – a close-knit collection of stars, usually outside the disk of our Galaxy, that formed many billions of years ago and contains many tens of thousands (or even more than a million) stars.

Planetary nebula – this is a short-lived stage in the life of stars having masses not too different from the Sun. Near the end of the nuclear fusion stage of such a star, gas is expelled in winds from the dying star, with these winds sometimes expelling more gas near the star's equator, and also interacting (e.g. fast winds catching up to slow winds), making interesting patterns in the gas that we see as different shapes to planetary nebulae. The gas is ionized by ultraviolet radiation from the central star, making the gas visible when electrons recombine with ions.

Emission nebula – this is a region where an interstellar gas cloud has been ionized by young, hot stars near or in these clouds. The clouds are mostly made of ionized hydrogen, but small amounts of other ionized atoms, such as oxygen, also emit light (for example, doubly ionized oxygen, or OIII, emits light at a particular wavelength that is easier to see with a special OIII filter that only lets this wavelength through).

Reflection nebula – this is a region where light from stars is scattered off dust in an interstellar cloud.

Elliptical galaxy – as the name suggests, these are galaxies with the shape of an ellipsoid (although many are not far from being spherical in shape).

Lenticular galaxy – lenticular means "lens-shaped" and this is the shape of these galaxies i.e. they are shaped like a convex lens and have a disk in their central plane (but this disk lacks spiral arms). Some have a bar in the disk, and this is noted by the classification "barred lenticular galaxy".

Spiral galaxy – these are galaxies with a disk shape that contains spiral arms within it. Those with a bar in the disk are indicated as "barred spiral galaxy". A galaxy is said to be "early-type" if it is an elliptical or lenticular galaxy, or is a spiral galaxy with relatively tightly wound spiral arms and a large central bulge (making it an "early-type" spiral). "Late-type" spirals have less tightly wound arms and a very small bulge compared to an extended disk.

Irregular galaxy – these are galaxies with no obvious rotational symmetry.

Supernova remnant – the visible remains of a supernova.

Asterism – a pattern of physically unrelated stars on the sky.

R.A., Dec.: The right ascension (R.A.) and declination (Dec.) coordinates of the object (Equinox 2000). R.A. and Dec. are analogous to longitude and latitude but refer to the object's position in the sky. R.A. is different from longitude, however, since it is measured in hours and minutes instead of degrees like declination, with 1 hour R.A. being 15° (so that 24 hours R.A. makes 360°), while 1 minute R.A. is 1/4° or 15′ angular measure. Note that 00h 00.0m R.A. runs through the east side of the Great Square of Pegasus.

Approximate date of transit at local midnight: This gives the approximate date on which the object is highest in the sky at midnight (and so is best positioned for viewing at midnight). The time of transit moves two hours earlier every month. For this reason, an object will transit at 10 p.m. approximately one month later than the date given and will transit at 8 p.m. two months later than the date given. For example, the entry for the Crab Nebula (M 1) gives "transit at local midnight" as December 24. Thus, at midnight on Christmas Eve, the Crab Nebula will be approximately due south for northern temperate zone observers, but will be due south at 10 p.m. around January 24 and at 8 p.m. around February 24.

The dates given assume daylight savings time is in effect for all of April through the end of October, unless otherwise specified. While this will be incorrect near the start of April and the end of October, the dates that are incorrect will vary from year to year and so correction for this is left to the reader. In addition, because the time of transit will depend on the observer's location within their time zone, variations by as much as two weeks from the date given can be expected (but are of little practical consequence).

Distance: Most astronomical distances are not known accurately, so that the distances listed here must be considered as approximate values. Most distances are accurate at best to 10%, but inaccuracies of up to a factor of two do occur (for example with planetary nebulae, whose distances are notoriously inaccurate). For example, if an object is listed as being at a distance of 5 thousand light years, then its distance is probably not known to an accuracy of better than ±0.5 thousand light years in the best cases, but in the worst cases its distance may only be known to lie in the range 2.5 thousand–10 thousand light years.

Age: The ages of most deep-sky objects are also not known to high accuracy, so that the values given are approximate values only. Inaccuracies are similar to those given under distance.

Apparent size: These are the approximate dimensions of the object in the sky as it would appear in a telescope. It should be noted that these dimensions will depend on the aperture of the telescope in many cases, since larger-diameter telescopes will often see larger dimensions for the object than smaller telescopes (since dimmer regions can be seen with larger telescopes). As a result, the given dimensions are approximate values. For reference, the apparent sizes of several commonly observed deep-sky objects are given in Table 1.

Table 1 Apparent sizes of several commonly observed objects	
Object	Apparent size
M 45 (Pleiades)	≈2°
M 42 (Orion Nebula)	≈1°
M 81 (spiral galazy in Ursa Major)	25′ × 12′
M 11 (Wild Duck Cluster)	14′
M 49 (brightest galaxy in Virgo cluster)	≈10′
M 27 (Dumbbell Nebula)	6′
M 57 (Ring Nebula)	≈1′
Separation of stars in the double star ε Lyrae (the "Double–Double")	2.3″, 2.6″

Magnitude: This is the total, integrated, visual magnitude of the object i.e. if all the visual light from the object were emitted from a point source (a star), this would be the magnitude of that point source. For diffuse objects it can be difficult to use total, integrated magnitudes to guess how bright the object would appear in the eyepiece. This is because what we see is the object's light spread out over the entire area of the object (the effect is similar to looking at a star in the eyepiece when the view is out of focus – the star appears much dimmer when out of focus even though the same amount of total light is reaching the eye from the star). The alternative is to use surface brightness, which has units of magnitude per unit area, but this requires knowing the area of the object and this has not been accurately measured for many deep-sky objects. Thus, only the total, integrated visual magnitude is given. It should be noted that the visual magnitude of some objects is somewhat uncertain, and for these objects the magnitude is given as an integer.

Sky Atlas 2000.0 chart: This gives the chart that the object can be found on in the star atlas *Sky Atlas 2000.0*, by W. Tirion & R.W. Sinnott, 2nd edition, Sky Publishing Corporation (Cambridge, US) and Cambridge University Press (Cambridge, UK), 1998.

Herald–Bobroff chart: This gives the sky chart that the object can be found on in the star atlas *Herald–Bobroff Astroatlas*, by D. Herald and P. Bobroff, HB Publications, Canberra, 1994.

Other Facts: Various interesting astrophysical facts about the object are given under this heading. Since a number of these refer to the relative location of objects within the Milky Way, a simplified schematic of the approximate basic dimensions of our Galaxy is shown in Figure 1. Our Galaxy is thought to be a barred spiral galaxy. Most of the luminous matter in our Galaxy is contained in the central bulge and thin disk, the latter being about 2 thousand light years in thickness and 1 hundred thousand light years in diameter. The bulge is a flattened spheroid with a major axis diameter of about 15 thousand light years and minor axis of about 6 thousand light years. At the center of the bulge lies what is thought to be a giant black hole with a mass of about 2.6 million Suns and a diameter that is no larger than that of our Solar System.

Unlike how it is shown in Figure 1, the Galactic disk does not end abruptly at its edges. Rather, it trails off over some distance, making its exact dimensions somewhat difficult to define. Most of the disk revolves at the same speed (a little over 2 hundred km/s). Our Sun lies in the disk close to the Galactic central plane at a distance of about 25 thousand light years from the center, as indicated in Figure 1, and has a period of about 1/4 billion years in its orbit around the Galaxy. The disk consists mostly of young and intermediate age ("Population I") stars that are typically between one million and 10 billion years old. Although not shown in Figure 1, the entire disk and bulge are embedded within a giant spherical region, termed the "halo",

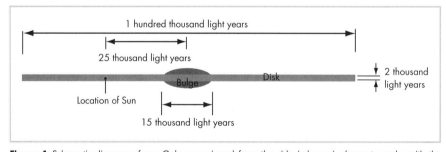

Figure 1 Schematic diagram of our Galaxy as viewed from the side (edge-on), drawn to scale, with the location of the Sun as indicated.

that may contain extensive amounts of dark matter extending out to a diameter of several hundred thousand light years. Within the halo lie many of the Galaxy's globular clusters. In contrast to the disk, the halo consists of old ("Population II") stars that are typically at least 10 billion years old. The total mass of the Milky Way (both dark and luminous matter) is somewhat uncertain but thought to be about 3/4 trillion Suns, with luminous stars making up less than a few hundred billion Suns of this mass.

For a number of galaxies, an inclination angle is mentioned in the text. This is the angle subtended by our line-of-sight and an axis perpendicular to the galaxy's central plane, as shown in Figure 2. The higher the inclination angle, the more flattened a disk galaxy appears to us, as shown in Figure 2 for several inclination angles. The inclination angle is defined so that it is always between 0° and 90°, so that no distinction is made between views from "above" or "below" a galaxy's central plane. For disk galaxies, the inclination angle can be approximated as arccos (minor axis/major axis), where the minor and major axes dimensions are listed under the object's "Apparent size" entry.

Number convention: Throughout the book, billion indicates 10^9 and trillion indicates 10^{12}, which is their common usage in North America and not to be confused with the usage of these words in the UK and other parts of the world.

In addition, a convention is adopted whereby natural numbers having two or more trailing zero digits have been written out in words e.g. 300,000 is written as 3 hundred thousand. The author has found this convention eases the reading of such numbers under dimly lit conditions (as occurs while

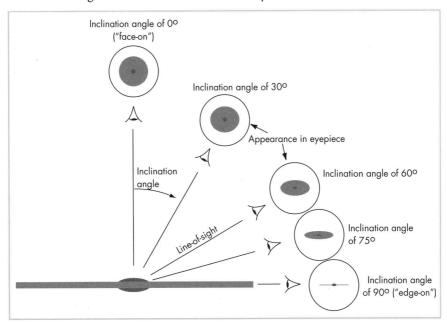

Figure 2 The concept of inclination angle is illustrated for a disk galaxy, with the appearance in the eyepiece shown at various inclination angles.

observing). In addition, it allows a quicker internalization of the size of such numbers than scientific notation, while avoiding the use of more significant digits than is warranted.

References: The information in this book was culled from several thousand archival journal papers, many of which were located using NASA's Astrophysics Data System. Additional sources of information included the following: NASA/IPAC Extragalactic Database (NED) which is operated by the Jet Propulsion Laboratory California Institute of Technology, under contract with the National Aeronautics and Space Administration; SIMBAD, operated by the Centre de Données astronomiques de Strasbourg; WEBDA, a web site devoted to stellar open clusters, developed and maintained by Jean-Claude Mermilliod, Institute of Astronomy of the University of Lausanne (Switzerland); the Revised New General Catalogue and Index Catalogue compilation of Wolfgang Steinicke; the Catalog of Parameters for Milky Way Globular Clusters, compiled by William E. Harris, McMaster University; the *Observer's Handbook*, edited by Rajiv Gupta, published yearly by the Royal Astronomical Society of Canada; the Innsbruck Data Base of Galactic Planetary Nebulae; the Saguaro Astronomy Club Database; TheSky Astronomy Software by Software Bisque; the *Encyclopedia of Astronomy and Astrophysics*, ed. P. Murdin, Institute of Physics, 2000; the Astronomical League's Herschel 400 list; William Herschel's original articles (Phil. Trans. Royal Soc. London vol. 76:457–499, 1786; vol. 79:212–255, 1789; vol. 92:477–528, 1802) and *Burnham's Celestial Handbook: An Observer's Guide to the Universe Beyond the Solar System*, R. B. Burnham, Dover, 1978.

2

The Messier Objects

M 1 (NGC 1952)

Constellation	Object type	RA, Dec	Approx. transit date at local midnight	Distance
Taurus	Supernova remnant	05h 34.5m, +22° 01′	December 24	6.5 thousand light years
Age	**Apparent size**	**Magnitude**	**Sky Atlas 2000.0 chart**	**Herald–Bobroff chart**
1 thousand years (created by a supernova in 1054 AD)	8′ × 4′	8.4	5	C35

Nicknamed the "Crab Nebula", this is one of the most well-studied objects in the sky. At its center is a pulsar, which is a rotating neutron star (about 10–15 km in diameter with a mass somewhat greater than the Sun) with a strong magnetic field that emits a narrow beam of radio emission. The Crab pulsar rotates about 30 times a second (i.e. its period is 33 milliseconds), with its beacon pointing at us once each rotation (like a very rapidly rotating lighthouse beam). Of the more than 1 thousand known radio pulsars, this is one of only five whose pulses are visible in optical wavelengths (with professional telescopes). Pulsars are the remains of a star that went supernova. For the Crab pulsar, the progenitor star is thought to have had a mass of 8–13 times that of our Sun (its exact mass is uncertain). The Crab Nebula constitutes the remnants of the material ejected by the supernova event of this star, with about five solar masses worth of material being present in the luminous portion of the nebula. The filaments in the nebula are moving outward at over a thousand km/s.

M 2 (NGC 7089)

Constellation	Object type	RA, Dec	Approx. transit date at local midnight	Distance
Aquarius	Globular cluster	21h 33.5m, −00° 49′	September 8	40 thousand light years
Age	**Apparent size**	**Magnitude**	**Sky Atlas 2000.0 chart**	**Herald–Bobroff chart**
12–14 billion years	11.7′	6.5	17	C41

Its mass is about 9 hundred thousand Suns, but many of these stars are more massive than the Sun so that the total number of stars is about 150 thousand. Its size of 11.7′ gives it a diameter of about 130 light years. It lies in the halo of our Galaxy. The halo is the region outside the spiral disk and bulge of our Galaxy, extending out as a sphere with a radius of perhaps six times that of the spiral disk region and containing most of the Galaxy's dark matter. M 2 orbits the Galaxy independently of the Galactic disk on an inclined orbit that wanders out over a hundred thousand light years from the Galactic center and then approaches within a few tens of thousands of light years of the Galactic center, taking the better part of a billion years to complete one revolution around the Galaxy.

M 3 (NGC 5272)

Constellation	Object type	RA, Dec	Approx. transit date at local midnight	Distance
Canes Venatici	Globular cluster	13h 42.2m, +28° 23′	May 11	30 thousand light years
Age	**Apparent size**	**Magnitude**	**Sky Atlas 2000.0 chart**	**Herald–Bobroff chart**
12–14 billion years	18.6′	6.2	7	C29

Its mass is nearly 8 hundred thousand Suns. Its size of 18.6′ corresponds to a diameter of about 160 light years. It orbits the Galaxy on a precessing elliptical path that is highly inclined with the plane of our Galaxy and quite eccentric (minor axis to major axis ratio of 0.4). It takes about 3 hundred million years to make one revolution of the Galaxy, never straying farther than about 50 thousand light years from the Galactic center (we are about 25 thousand light years from the Galactic center, although we stay in the disk of our Galaxy and M 3 does not). M 3 never approaches closer than about 15 thousand light years from the Galactic center, making it an inner halo cluster (so called since it doesn't travel too far out in the halo – see M 2 for the meaning of "halo").

M 4 (NGC 6121)

Constellation	Object type	RA, Dec	Approx. transit date at local midnight	Distance
Scorpius	Globular cluster	16h 23.6m, −26° 32′	June 21	7 thousand light years
Age	**Apparent size**	**Magnitude**	**Sky Atlas 2000.0 chart**	**Herald–Bobroff chart**
12–14 billion years	22.8′	5.6	22	C63/D13

This is the nearest globular cluster to us. Its size of 22.8′ corresponds to a diameter of about 50 light years. It has a mass of about 2 hundred thousand Suns. It is the only globular cluster known to harbor a triple star system. Even more unusual is that this is the only known triple star system that contains a pulsar (see M 1 for explanation of pulsar). This triple star consists of a neutron star with a mass about 1.4 times that of our Sun, a white dwarf with a mass about 30% that of our Sun, and a "planet" with < 1% the mass of our Sun. This planet may have been stolen from another star–planet system by the neutron star in the past few million years. The cluster lies toward the Galactic center, within roughly 2 thousand light years of the Galactic central plane, so that interstellar material in the disk of our Galaxy blocks out some of its light and makes it dimmer (by a few magnitudes) than it would otherwise appear.

M 5 (NGC 5904)

Constellation	Object type	RA, Dec	Approx. transit date at local midnight	Distance
Serpens Caput	Globular cluster	15h 18.6m, +02° 05′	June 5	24 thousand light years
Age	**Apparent size**	**Magnitude**	**Sky Atlas 2000.0 chart**	**Herald–Bobroff chart**
12–14 billion years	19.9′	5.7	15	C45

Its mass is over 8 hundred thousand Suns and its diameter is about 140 light years. Like most globular clusters it does not orbit our Galaxy with the Galactic disk as we do. Instead it follows an orbit that takes it out as far as 150 thousand light years from the Galactic center and then back in as close as about 10 thousand light years, on a path highly inclined to the Galactic disk, taking almost a billion years to make one revolution around our Galaxy (compare to the 1/4 billion years our Sun takes to make one Galactic revolution). M 5 currently sits about 20 thousand light years from the Galactic center, which is much closer than its average orbital distance.

M 6 (NGC 6405)

Constellation	Object type	RA, Dec	Approx. transit date at local midnight	Distance
Scorpius	Open cluster	17h 40.3m, −32° 15′	July 10	1.6 thousand light years
Age	**Apparent size**	**Magnitude**	**Sky Atlas 2000.0 chart**	**Herald–Bobroff chart**
80 million years	20′	4.0	22	C62/D12

Nicknamed the "Butterfly Cluster" due to the shape of its apparent outline. It contains several chemically peculiar stars ("CP2" stars – see NGC 7243 and NGC 2169 for explanation). It lies in the direction of the center of our Galaxy, less than twenty light years below the Galactic central plane. Its size of 20′ corresponds to a diameter of about 10 light years. Professional telescopic studies have counted over 3 hundred stars belonging to this cluster (only a small fraction of which are visible in amateur telescopes).

M 7 (NGC 6475)

Constellation	Object type	RA, Dec	Approx. transit date at local midnight	Distance
Scorpius	Open cluster	17h 53.9m, −34° 48′	July 14	1 thousand light years
Age	**Apparent size**	**Magnitude**	**Sky Atlas 2000.0 chart**	**Herald–Bobroff chart**
2 hundred million years	80′	3.3	22	C62/D12

First mentioned by Ptolemy over 2 thousand years ago, it has a diameter of about 20 light years. Professional telescopic studies find that nearly 100% of the stars in M 7 are binary stars, an inordinately high frequency of binaries compared to the Galactic field (where > 50% of main-sequence stars are binaries). Professional telescopic studies have counted about a hundred stars belonging to this cluster. Like all open clusters, a good number of the stars in the field of this cluster (called "field stars") do not belong to the cluster. Instead, they just happen to lie in the line-of-sight of the cluster but are actually much closer or farther away from us than the cluster. This "contamination" by field stars can be seen in Figure 6475 (overleaf) (see M 39 for further discussion of this).

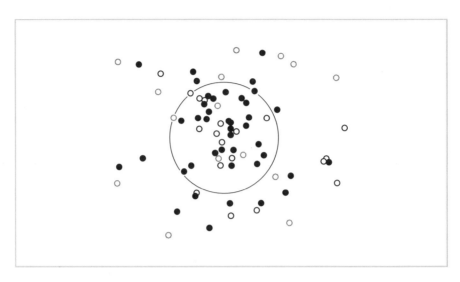

Figure 6475 The brighter stars (< mag. 10) in the region of M 7 are shown. The large circle is shown to give a length scale and has a diameter of 60′. Stars shown by a solid circle can be considered as belonging to M 7 since they have a greater than 90% chance of belonging to the cluster. However, stars represented by faint open circles probably do not belong to M 7, since they have a less than 20% change of belonging to the cluster. Bold open circles may or may not belong to the cluster, since they have a membership probability between 20% and 90%. Reprinted from F. Gieseking (Astron. Astrophys. Suppl. Ser. 61:75–81, 1985) with permission.

M 8 (NGC 6523)

Constellation	Object type	RA, Dec	Approx. transit date at local midnight	Distance
Sagittarius	Emission nebula	18h 03.7m, −24° 23′	July 16	5 thousand light years
Age	**Apparent size**	**Magnitude**	**Sky Atlas 2000.0 chart**	**Herald–Bobroff chart**
	45′ × 30′	5.0	22	C61/D11

Nicknamed the "Lagoon Nebula". The open cluster NGC 6530 is embedded within this nebula. The stars in this cluster formed from the nebula in the past 2 million years or so. The nebula itself is a large ionized hydrogen (or HII) region, within which young O and B type stars associated with the cluster are thought to still be triggering star formation. M 8 is actually only a small "blister" on the surface of a giant molecular gas cloud that lies behind M 8. The nebula is ionized (and thus made visible) largely by just three stars (HD 165052, 9 Sgr, and HD 164740) as shown in Figure 6523. HD 164740 is a double star with a separation of 39″. It is responsible for ionizing the brightest part of the nebula, a 15″ × 30″ patch in the center of the nebula. This patch is called the "Hourglass" and is only about 10 thousand years old. The Hourglass is on the backside of the nebula, right on the edge of the giant molecular cloud from which NGC 6530 has formed.

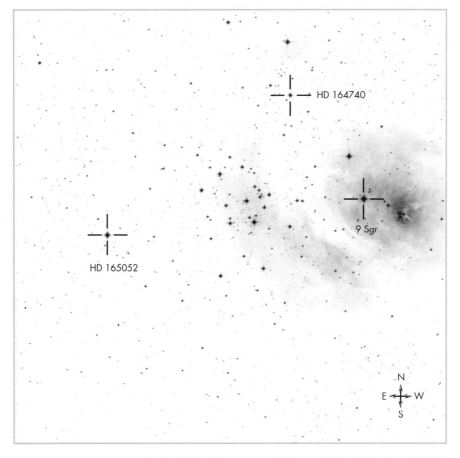

Figure 6523 The region east of and including the emission nebula M 8 (NGC 6523) is shown (with the open cluster NGC 6530 in the middle of the figure), with the three stars responsible for ionizing M 8 marked with cross-hairs. The area shown is 30′ × 30′. From the Digitized Sky Survey (Space Telescope Science Institute) based on photographic data of the National Geographic Society – Palomar Observatory Sky Survey (POSS I).

M 9 (NGC 6333)

Constellation	Object type	RA, Dec	Approx. transit date at local midnight	Distance
Ophiuchus	Globular cluster	17h 19.2m, −18° 31′	July 5	27 thousand light years
Age	**Apparent size**	**Magnitude**	**Sky Atlas 2000.0 chart**	**Herald– Bobroff chart**
12–14 billion years	5.5′	7.7	15	C62/D13

Its mass is about 3 hundred thousand Suns. Its size of 5.5′ gives it a diameter of about 40 thousand light years. It belongs to the central bulge of our Galaxy (i.e. the central, spherically shaped region within about 15 thousand light years of the Galactic center) and lies a few thousand light years nearly directly below the Galactic center. This is a metal-poor bulge cluster (metal-poor meaning it has a low abundance of elements heavier than helium) and so may have formed in the earliest star-forming period of our Galaxy (see NGC 6287).

M 10 (NGC 6254)

Constellation	Object type	RA, Dec	Approx. transit date at local midnight	Distance
Ophiuchus	Globular cluster	16h 57.1m, –04° 06′	June 30	15 thousand light years
Age	**Apparent size**	**Magnitude**	**Sky Atlas 2000.0 chart**	**Herald–Bobroff chart**
	12.2′	6.6	15	C44

Its mass is a little over 2 hundred thousand Suns. Its size of 12.2′ corresponds to a diameter of about 50 light years. This cluster stays within several thousand light years of the Galactic central plane and has a velocity close to that of the material in the Galactic disk. This is quite unusual for a "metal-poor" cluster like this one ("metal-poor" meaning it has low amounts of elements heavier than helium), since stars in the Galactic disk tend to have "metal" that was scattered by previous supernovae. The origin of this cluster is thus uncertain. Models predict this cluster will undergo core collapse in perhaps about 10 billion years, a process that causes stars near the cluster center to confine themselves to an inordinately small region of space like a swarm of angry bees suddenly placed into a small container – see NGC 6284. Core collapse will be followed by destruction of the cluster by gravitational interactions in about twice this time.

M 11 (NGC 6705)

Constellation	Object type	RA, Dec	Approx. transit date at local midnight	Distance
Scutum	Open cluster	18h 51.1m, –06° 16′	July 29	6 thousand light years
Age	**Apparent size**	**Magnitude**	**Sky Atlas 2000.0 chart**	**Herald–Bobroff chart**
250 million years	14′	5.8	16	C43/D10

Nicknamed the "Wild Duck Cluster" after the V-shaped outline (pointed east) that some of its brighter members make. Its mass is several thousand Suns, with 5 hundred members brighter than mag. 14. Its size of 14′ corresponds to a diameter of about 25 light years. A person in the middle would see a night sky with several hundred first mag. stars, each separated by < 1 light year. This is nearly as dense as some globular clusters. A significant number of field stars are present as foreground/background stars (see M 39).

M 12 (NGC 6218)

Constellation	Object type	RA, Dec	Approx. transit date at local midnight	Distance
Ophiuchus	Globular cluster	16h 47.2m, –01° 57′	June 27	22 thousand light years
Age	**Apparent size**	**Magnitude**	**Sky Atlas 2000.0 chart**	**Herald–Bobroff chart**
12–14 billion years	12.2′	6.7	15	C44

Its mass is about 1/4 million Suns. Its size of 12.2′ corresponds to a diameter of about 80 light years. Like M 3 and M 10, this is an inner halo cluster, so called since it doesn't travel too far out in the halo – see M 2 for the meaning of "halo". M 12 never strays farther than about 20 thousand light years from the Galactic center on an orbit inclined to the Galactic central plane by 33° or so. M 12 takes about 130 million years to complete one revolution around the Galaxy, having just crossed the Galactic central plane a few million years ago (lying 2 thousand light years below it) on its way to a maximum excursion of a little under 10 thousand light years below the Galactic plane. (Contrast this with the Sun, which stays in the 2 thousand light year thick disk of the Galaxy and is currently about 50 light years above the Galactic central plane, having crossed the central plane about 2 million years ago, on its way to a maximum excursion of about 250 light years above the Galactic central plane.)

M 13 (NGC 6205)

Constellation	Object type	RA, Dec	Approx. transit date at local midnight	Distance
Hercules	Globular cluster	16h 41.7m, +36° 28′	June 26	25 thousand light years
Age	**Apparent size**	**Magnitude**	**Sky Atlas 2000.0 chart**	**Herald–Bobroff chart**
12–14 billion years	17′	5.8	8	C26

Nicknamed the "Hercules Cluster". Its 17′ size corresponds to a diameter of about 120 light years. It contains about 6 hundred thousand solar masses. It orbits the Galaxy independently of the material in the Galactic disk on an inclined orbit that travels out to about 80 thousand light years from the Galactic center but approaches within 20 thousand light years of the Galactic center, taking a 1/2 billion or so years to complete one revolution.

M 14 (NGC 6402)

Constellation	Object type	RA, Dec	Approx. transit date at local midnight	Distance
Ophiuchus	Globular cluster	17h 37.6m, –03° 15′	July 10	30 thousand light years
Age	**Apparent size**	**Magnitude**	**Sky Atlas 2000.0 chart**	**Herald–Bobroff chart**
12–14 billion years	6.7′	7.6	15	C44

Its mass is about 1.2 million times that of our Sun. Its size of 6.7′ corresponds to a diameter of about 60 light years. It is located in the central bulge of our Galaxy (i.e. the central, spherically shaped region) and is relatively lacking in elements heavier than helium (i.e. "metals"), so that it may have formed in one of the earliest star-forming periods of our Galaxy (see NGC 6287). It lies about 8 thousand light years above the Galactic central plane about 13 thousand light years from the Galactic center. It was only the second globular cluster (after M 80) to have a nova discovered in it.

M 15 (NGC 7078)

Constellation	Object type	RA, Dec	Approx. transit date at local midnight	Distance
Pegasus	Globular cluster	21h 30.0m, +12° 10′	September 7	34 thousand light years
Age	**Apparent size**	**Magnitude**	**Sky Atlas 2000.0 chart**	**Herald–Bobroff chart**
12–14 billion years	12.3′	6.2	17	C41

Its mass is nearly a million Suns. Its size of 12.3′ corresponds to a diameter of 120 light years. It is a halo cluster (see M 2 for the meaning of "halo"), but never travels farther than about 45 thousand light years from the Galactic center on a path that is inclined by about 40° from the Galactic disk. It revolves once around the Galaxy every 1/4 billion years or so in a prograde orbit (like most globular clusters), meaning it revolves about the Galaxy in the same direction as the Galaxy's own rotation. The cluster is core collapsed (see NGC 6284 for explanation) and has one of the most concentrated centers (with more than 30 stars per square arcsecond in professional telescopes). It has been proposed that the cluster may contain a central black hole (with a mass of several thousand Suns), but this remains uncertain. M 15 does contain a planetary nebula (Pease 1, mag. 13), one of only four globular clusters that share this distinction and the easiest planetary of the four to find in amateur telescopes (but recommended for a 12-inch or larger telescope, and requiring a detailed map of the field, optional nebula filter, and patience to discern Pease 1 among the myriad stars near it).

M 16 (NGC 6611)

Constellation	Object type	RA, Dec	Approx. transit date at local midnight	Distance
Serpens Cauda	Open cluster with nebulosity	18h 18.8m, −13° 47′	July 20	6 thousand light years
Age	**Apparent size**	**Magnitude**	**Sky Atlas 2000.0 chart**	**Herald–Bobroff chart**
6 million years	7′	6.0	16	C61/D11

This open cluster is embedded in a gas cloud from which the cluster formed and in which star formation is still going on. Young, hot stars in the cluster are ionizing the surrounding hydrogen gas cloud (making it a so-called HII region), thereby making it fluoresce as the emission nebula IC 4703 which is nicknamed the "Eagle Nebula" or the "Star Queen Nebula" after the appearance of part of this nebula in professional telescopic images and photographs. This namesake region is well known from the Hubble Space Telescope's "Pillars of Creation" photo of a 2′ × 2′ or so portion of it (which lies just SE of the open cluster's most concentrated area). This photo shows three large pillars (looking like "elephant trunks" or hoodoos) aligned in a SE–NW direction. The pillars are regions of dark molecular gas and dust that are being "eroded" by intense radiation from stars to their NW. Professional telescopic studies have counted almost 4 hundred stars as members of the open cluster (but only a fraction of these are visible in amateur telescopes and are easily confused with field stars – see M 39). The cluster's size of 7′ corresponds to a diameter of about 10 light years.

M 17 (NGC 6618)

Constellation	Object type	RA, Dec	Approx. transit date at local midnight	Distance
Sagittarius	Emission nebula + open cluster	18h 20.8m, −16° 11′	July 21	7 thousand light years
Age	Apparent size	Magnitude	Sky Atlas 2000.0 chart	Herald–Bobroff chart
1 million years	11′	6.0	16	C61/D11

The appearance of this nebula in amateur telescopes leads to its various nicknames ("Swan Nebula", "Omega Nebula", among others). Its size of 11′ corresponds to a diameter of about 20 light years. Like M 8 and M 16, the nebula fluoresces due to an embedded open cluster (containing 35 or so stars) that formed from the nebula. However, the stars in the open cluster are so heavily obscured by intervening gas and dust that only five of them have magnitudes brighter than 14 (with only two brighter than magnitude 10), making its appearance as a true "cluster" of stars essentially nonexistent in the eyepiece of an amateur telescope.

M 18 (NGC 6613)

Constellation	Object type	RA, Dec	Approx. transit date at local midnight	Distance
Sagittarius	Open cluster	18h 19.9m, −17° 08′	July 21	5 thousand light years
Age	Apparent size	Magnitude	Sky Atlas 2000.0 chart	Herald–Bobroff chart
30 million years	9′	6.9	16	C61/D11

This sparse cluster contains about 20 members, has a diameter somewhat over 10 light years and has not been well studied.

M 19 (NGC 6273)

Constellation	Object type	RA, Dec	Approx. transit date at local midnight	Distance
Ophiuchus	Globular cluster	17h 02.6m, −26° 16′	July 1	30 thousand light years
Age	Apparent size	Magnitude	Sky Atlas 2000.0 chart	Herald–Bobroff chart
12–14 billion years	12.3′	6.2	22	C62/D13

Its mass is about 1.5 million Suns. This is the most flattened globular cluster known in our Galaxy (its oval shape is apparent in amateur telescopes). It gives off the second most light of all the NGC globular clusters in our Galaxy, after ω Centauri (NGC 5139). It belongs to the central bulge of our Galaxy (i.e. the central, spherically shaped region), lying a few thousand light years nearly directly above the Galactic center. It is a metal-poor bulge cluster (metal-poor meaning it has a low abundance of elements heavier than helium) and so may have formed in the earliest star-forming period of our Galaxy (see NGC 6287).

M 20 (NGC 6514)

Constellation	Object type	RA, Dec	Approx. transit date at local midnight	Distance
Sagittarius	Emission and reflection nebula	18h 02.7m, –22° 58'	July 16	5 thousand light years
Age	Apparent size	Magnitude	Sky Atlas 2000.0 chart	Herald–Bobroff chart
3 hundred thousand years	28'	6.3	22	C61/D11

Light from very young stars that formed out of the surrounding gas ionize hydrogen in the nebula causing it to glow (making the nebula a so-called HII region, although other elements are present, in particular oxygen, since the nebula benefits from a filter that lets in light from doubly ionized oxygen i.e. OIII). This is one of the youngest HII regions known, and is thought to be in a "pre-Orion-nebula" state, where young stars are violently ejecting matter, and protostars with jets are interacting with the nebula. The nebula is "lit up" mostly by the bright star in the middle of the nebula (HD 164492 or ADS 10991, mag. 7). ADS 10991 is a multiple star whose two brightest components have a separation of 10.6" and position angle of 212°. It is found to consist of seven stars in professional telescopic studies. Dust grains are also present in the nebula (in the northern regions near the bright mag. 7.5 star there) which scatter the starlight, so that this nebula consists of both a reflection nebula (in the north) and an emission nebula (in the south). The nickname "Trifid Nebula" refers to the emission nebula, and is derived from Latin (trifidus, meaning "split into three"), which refers to its three lobes, which are separated by lanes of dust grains in the nebula that block its light from us. The entire nebula's size of 28' corresponds to a diameter of about 40 light years.

M 21 (NGC 6531)

Constellation	Object type	RA, Dec	Approx. transit date at local midnight	Distance
Sagittarius	Open cluster	18h 04.2m, –22° 29'	July 16	4 thousand light years
Age	Apparent size	Magnitude	Sky Atlas 2000.0 chart	Herald–Bobroff chart
10 million years	13'	5.9	22	C61/D11

This cluster lies very close to the Galactic central plane (being a few tens of light years below it). Of the 1.5 thousand or so open clusters in our Galaxy this is one of many that have not been very well studied.

M 22 (NGC 6656)

Constellation	Object type	RA, Dec	Approx. transit date at local midnight	Distance
Sagittarius	Globular cluster	18h 36.4m, −23° 54′	July 25	10 thousand light years
Age	**Apparent size**	**Magnitude**	**Sky Atlas 2000.0 chart**	**Herald–Bobroff chart**
12–14 billion years	17′	5.1	22	C61/D11

Its mass is about 1/2 million Suns. Along with M 15, this is one of only four globular clusters that contains a planetary nebula, labeled GJJC 1. However, at mag. 15 and lying near the core of the cluster, finding GJJC 1 in an amateur telescope is a challenge this author has not attempted (the one in M 15 is easier to find – see M 15). Recent professional telescopic observations of several stars in the bulge of our Galaxy on the other side of M 22 lying in the same field as M 22 from our line-of-sight ("field stars") showed transient brightening (by gravitational microlensing). One explanation for this is that a cluster of dark objects containing perhaps a million MACHOs (massive compact halo objects, like brown dwarves and planets) may lie between M 22 and the bulge of our Galaxy. However, it has also been suggested that planets within M 22 could be responsible for the microlensing – a definitive explanation awaits future research. M 22 never strays too far from the Galactic disk in its orbit, staying within about 5 thousand light years of the Galactic central plane between about 30 thousand and 10 thousand light years from the Galactic center, orbiting the Galaxy once every 2 hundred million years or so.

M 23 (NGC 6494)

Constellation	Object type	RA, Dec	Approx. transit date at local midnight	Distance
Sagittarius	Open cluster	17h 57.1m, −18° 59′	July 15	2 thousand light years
Age	**Apparent size**	**Magnitude**	**Sky Atlas 2000.0 chart**	**Herald–Bobroff chart**
3 hundred million years	27′	5.5	15	C61/D11

Its size of 27′ corresponds to a diameter of about 15 light years. Professional telescopic studies count about 150 member stars that have less than a 10% chance of being "field stars" – see M 39.

M 24

Constellation	Object type	RA, Dec	Approx. transit date at local midnight	Distance
Sagittarius	Star cloud	18h 16.5m, −18° 50′	July 15	12–16 thousand light years
Age	**Apparent size**	**Magnitude**	**Sky Atlas 2000.0 chart**	**Herald–Bobroff chart**
	1.6° × 0.6°	4.6	16	C61/D11

This is a patch of the Milky Way seen through a hole in the foreground interstellar dust that obscures the surrounding sky, making the patch appear as a cluster to Messier even though it is not a true cluster. It is nicknamed the "Small Sagittarius Star Cloud". The open cluster NGC 6603 (mag. 11, 12 thousand light years away, 2 hundred million years old) lies within M 24 (probably on the near side of the star cloud that makes up M 24) and is often incorrectly labeled as M 24 instead.

M 25 (IC 4725)

Constellation	Object type	RA, Dec	Approx. transit date at local midnight	Distance
Sagittarius	Open cluster	18h 31.8m, −19° 07′	July 24	2 thousand light years
Age	**Apparent size**	**Magnitude**	**Sky Atlas 2000.0 chart**	**Herald–Bobroff chart**
90 million years	29′	4.6	16	C61/D11

It has a diameter of about 20 light years and lies about 2 hundred light years below the Galactic central plane. It contains one Cepheid variable star (U Sgr, mag. 7, the right-hand star marked as U on Herald–Bobroff chart D11 – see NGC 7790 for explanation of Cepheid variables). It also contains six Be stars (see M 47 for explanation of Be stars).

M 26 (NGC 6694)

Constellation	Object type	RA, Dec	Approx. transit date at local midnight	Distance
Scutum	Open cluster	18h 45.2m, −09° 24′	July 27	5 thousand light years
Age	**Apparent size**	**Magnitude**	**Sky Atlas 2000.0 chart**	**Herald–Bobroff chart**
90 million years	15′	8.0	16	C43/D10

It has a diameter of about 20 light years. It lies about 250 light years below the Galactic central plane, which happens to be the furthest extent our Sun travels from the Galactic central plane (although the Sun currently sits about 50 light years above the Galactic central plane).

M 27 (NGC 6853)

Constellation	Object type	RA, Dec	Approx. transit date at local midnight	Distance
Vulpecula	Planetary nebula	19h 59.6m, +22° 43′	August 15	1 thousand light years
Age	**Apparent size**	**Magnitude**	**Sky Atlas 2000.0 chart**	**Herald–Bobroff chart**
	5.8′	7.3	8	C24/D09

Nicknamed the "Dumbbell Nebula", although a partially eaten apple might be a better description of its actual appearance in amateur telescopes. Its bipolar (bowtie-shaped) nature occurs in < 20% or so of all planetary nebulae. The shape of such planetary nebulae is thought to begin with an aging giant star giving off a large amount of gas in a "superwind" (travelling at 10 km/s, emitting 10^{-4} solar masses/year) that is concentrated near the star's equator. Once the core of the old star is eventually exposed, a hot, fast wind (1 thousand km/s, emitting 10^{-9} solar masses/year) slams into the previously emitted gas that is concentrated in the equatorial regions. This results in preferential expansion of this wind in the polar directions and the two familiar bipolar lobes. The lobes are ionized by short-wavelength, nonvisible radiation from the central star, and re-emit this radiation in visible wavelengths. The central star (mag. 13.8) lies at the narrowest part of the "bowtie" shape. The nebula is several thousand years old and still expanding (at several tens of km/s). In professional telescopes M 27 is found to have an elliptical shell surrounding it, as well as an internal elliptical shell, so that the structure of this nebula is quite complex.

M 28 (NGC 6626)

Constellation	Object type	RA, Dec	Approx. transit date at local midnight	Distance
Sagittarius	Globular cluster	18h 24.5m, −24° 52′	July 22	20 thousand light years
Age	**Apparent size**	**Magnitude**	**Sky Atlas 2000.0 chart**	**Herald–Bobroff chart**
	15′	6.8	22	C61/D11

Its mass is almost 1/2 million Suns. Like M 10, this cluster spends its time within a few thousand light years of the Galactic central plane and rotates approximately with the material in the Galactic disk. This is quite unusual for a "metal-poor" cluster like this one ("metal-poor" meaning it has low amounts of elements heavier than helium), since stars in the Galactic disk tend to have "metal" that was scattered by previous supernovae. The origin of this cluster is thus uncertain.

M 29 (NGC 6913)

Constellation	Object type	RA, Dec	Approx. transit date at local midnight	Distance
Cygnus	Open cluster	20h 23.9m, +38° 32′	August 21	3 thousand light years
Age	**Apparent size**	**Magnitude**	**Sky Atlas 2000.0 chart**	**Herald–Bobroff chart**
A few million years	7′	6.6	9	C24/D08

It has a diameter of less than 10 light years. It is a very young cluster (young enough that circumstellar disks associated with planet formation could still be present – see NGC 2362). However, it is heavily obscured by foreground dust that is very patchy (dimming some stars in the cluster by up to 5 magnitudes, but hardly dimming others) making studies on it difficult.

M 30 (NGC 7099)

Constellation	Object type	RA, Dec	Approx. transit date at local midnight	Distance
Capricornus	Globular cluster	21h 40.4m, −23° 11′	September 9	25 thousand light years
Age	**Apparent size**	**Magnitude**	**Sky Atlas 2000.0 chart**	**Herald–Bobroff chart**
12–14 billion years	9′	7.2	23	C59

It has a mass of about 3 hundred thousand Suns. Like most globular clusters, it orbits the Galaxy on a path that is inclined to the Galactic disk (by 50°), taking about 160 million years to complete one revolution around the Galaxy, never straying farther than about 25 thousand light years from the Galactic center, but never approaching closer than about 10 thousand light years to the Galactic center. The core of this cluster has "collapsed" (see NGC 6284), making its central region like a swarm of angry bees suddenly placed into a small container. Its size corresponds to a diameter of over a hundred light years. It contains the highest known globular cluster concentration of "blue straggler" stars. These are stars that are paradoxically far more blue and luminous than expected for reasons that remain unknown, but may involve the coalescence of stars (see NGC 6633), with almost 50 such stars known in this cluster.

M 31 (NGC 224)

Constellation	Object type	RA, Dec	Approx. transit date at local midnight	Distance
Andromeda	Spiral galaxy	00h 42.7m, +41° 16′	October 26 (Daylight Savings Time)	2.5 million light years
Age	Apparent size	Magnitude	Sky Atlas 2000.0 chart	Herald–Bobroff chart
	3.1° × 1.0°	3.4	4	C38

Nicknamed the "Andromeda Galaxy". This is the nearest large galaxy and is the most luminous member of our Local Group of galaxies (which contains about 40 galaxies in a radius of a few million light years). M 31 is thought to be roughly as massive as our Galaxy, but may be less massive (which would make our Galaxy the most massive member of the Local Group), although this remains uncertain. An apparent disk diameter of 3.1° corresponds to a diameter of 135 thousand light years, although the diameter visible in telescopes is larger than this (and depends on the aperture and observing conditions). Professional telescopes show it has perhaps as many as 15 dwarf spheroidal satellite galaxies, two of which [M 32 and NGC 205 (M 110)] are prominent in amateur telescopes. This is only a little more than the 11 such dwarf companions of our Galaxy. M 31 is very rich in globular clusters for a spiral galaxy, with nearly 5 hundred revealed in professional telescopes. The brightest of these globular clusters can be observed in amateur telescopes. Some of them are thought to have come from dwarf galaxies that M 31 accreted in past merger events. Professional telescopes find M 31 has a double nucleus containing a black hole with a mass of about 30 million Suns. The two nuclei are separated by 0.5″ and are thought to be part of an eccentric nuclear disk. The galaxy is a LINER ("low-ionization nuclear emission region" – see M 81/NGC 3031 for explanation).

M 32 (NGC 221)

Constellation	Object type	RA, Dec	Approx. transit date at local midnight	Distance
Andromeda	Elliptical galaxy	00h 42.7m, +40° 52′	October 26 (Daylight Savings Time)	2.4 million light years
Age	Apparent size	Magnitude	Sky Atlas 2000.0 chart	Herald–Bobroff chart
	8.5′ × 6.5′	8.1	4	C38

This is a dwarf satellite galaxy of M 31 and is M 31's closest companion. M 32 is a very unusual galaxy, referred to as a "compact elliptical" galaxy, having an inordinately bright and compact central core. Its origin remains uncertain, but one hypothesis suggests it was once a spiral galaxy that was stripped down to its bulge a few billion years ago by intense tidal interaction with M 31. A black hole with a mass of several million Suns is thought to be present in the center of M 32. M 32 has an optical diameter of about 6 thousand light years.

M 33 (NGC 598)

Constellation	Object type	RA, Dec	Approx. transit date at local midnight	Distance
Triangulum	Spiral galaxy	01h 33.9m, +30° 39'	October 24 (Standard Time)	2.8 million light years

Age	Apparent size	Magnitude	Sky Atlas 2000.0 chart	Herald–Bobroff chart
	68.7' × 41.6'	5.7	4	C38/D42

This is the third most luminous galaxy in our Local Group (which contains about 40 galaxies in a radius of a few million light years) after M 31 and the Milky Way. It is nicknamed the "Pinwheel Galaxy". It has an optical diameter of about 60 thousand light years (about half that of the Milky Way), which is about average for a spiral galaxy. Its luminous mass is about 10 billion Suns. It rotates clockwise from our viewpoint with a period of about 2 hundred million years. The size of its central black hole (if one exists) is at most about 3 thousand solar masses. NGC 604, the largest known ionized hydrogen region (1.5 thousand light years in diameter), belongs to this galaxy and is visible as a knot in large amateur telescopes near the NNE edge.

M 34 (NGC 1039)

Constellation	Object type	RA, Dec	Approx. transit date at local midnight	Distance
Perseus	Open cluster	02h 42.1m, +42° 46'	November 10	1.5 thousand light years

Age	Apparent size	Magnitude	Sky Atlas 2000.0 chart	Herald–Bobroff chart
250 million years	35'	5.2	4	C37/D41

It has a diameter of about 15 light years and was discovered in the middle of the seventeenth century by Giovanni Batista Hodierna. Its stars rotate at rates that are midway between those in the younger Pleiades cluster (100 million years old – see M 45) and the older Hyades cluster (6 hundred million years old). This is thought to be the result of rotational braking whose effect on rotation rates becomes more pronounced with age. Such braking is believed to be due to angular momentum loss via magnetic coupling to the chromosphere (i.e. the star's atmosphere outside the bright photosphere).

M 35 (NGC 2168)

Constellation	Object type	RA, Dec	Approx. transit date at local midnight	Distance
Gemini	Open cluster	06h 09.0m, +24° 21'	January 2	3 thousand light years

Age	Apparent size	Magnitude	Sky Atlas 2000.0 chart	Herald–Bobroff chart
2 hundred million years	28'	5.1	5	C34

It lies almost directly in the Galactic anticenter direction (i.e. directly outward from us in the opposite direction from the center of the Galaxy), about one hundred light years above the Galactic central plane. The open cluster NGC 2158 (see NGC 2158) lies only 24' SW, but is not near M 35 in space (NGC 2158 is roughly 13 thousand light years farther away). The total mass of M 35 is several thousand Suns.

M 36 (NGC 1960)

Constellation	Object type	RA, Dec	Approx. transit date at local midnight	Distance
Auriga	Open cluster	05h 36.3m, +34° 08′	December 25	4 thousand light years
Age	**Apparent size**	**Magnitude**	**Sky Atlas 2000.0 chart**	**Herald–Bobroff chart**
20 million years	12′	6.0	5	C35/D25

It lies about 70 light years above the Galactic central plane and has a diameter of about 15 light years. Its neighbors M 37 (3° 45′ ESE) and M 38 (2° 16′ NW) lie within a thousand light years of M 36. Although it is quite a young cluster, it is old enough that almost all its stars have had time to lose their youthful circumstellar disks (that formed when the stars collapsed from the surrounding gas and dust – see NGC 2362).

M 37 (NGC 2099)

Constellation	Object type	RA, Dec	Approx. transit date at local midnight	Distance
Auriga	Open cluster	05h 52.3m, +32° 33′	December 29	4 thousand light years
Age	**Apparent size**	**Magnitude**	**Sky Atlas 2000.0 chart**	**Herald–Bobroff chart**
4 hundred million years	24′	5.6	5	C35/D25

It has a diameter of about 30 light years. It lies close to M 38 and M 36 (see M 36). This is a rich cluster, with well over 2 thousand stars having been counted as members in professional telescopic studies.

M 38 (NGC 1912)

Constellation	Object type	RA, Dec	Approx. transit date at local midnight	Distance
Auriga	Open cluster	05h 28.7m, +35° 51′	December 23	4 thousand light years
Age	**Apparent size**	**Magnitude**	**Sky Atlas 2000.0 chart**	**Herald–Bobroff chart**
3 hundred million years	21′	6.4	5	C35/D25

It has a diameter of about 25 light years and lies about 50 light years above the Galactic central plane. It is physically close to M 37 and M 36 (see M 36). NGC 1907, 32′ SSW, has a projected distance (i.e. the distance perpendicular to our line-of-sight) of about 40 light years from M 38. It has been proposed that the two clusters formed from the same cloud (since they share a reasonably similar age, similar amounts of elements heavier than helium, and similar velocities relative to us), making them a double cluster if proven true.

M 39 (NGC 7092)

Constellation	Object type	RA, Dec	Approx. transit date at local midnight	Distance
Cygnus	Open cluster	21h 31.7m, +48° 26′	September 7	1 thousand light years
Age	**Apparent size**	**Magnitude**	**Sky Atlas 2000.0 chart**	**Herald–Bobroff chart**
3 hundred million years	32′	4.6	9	C05/D07

It has a diameter of about 10 light years. This cluster lies in a rich field that is "contaminated" with field stars (i.e. stars that happen to lie along the same line-of-sight but which are foreground or background stars) from the Milky Way. This "contamination" worsens the fainter the stars that are being considered. For example, about 80–90% of the magnitude 8–10 stars are true cluster members, but only about 20% of the magnitude 11 stars in this cluster are actual cluster members (the rest being field stars), while fewer than 10% of the mag. 12 stars are true cluster members. Distinguishing field stars from true open cluster members requires professional telescopic studies, but the fact that one is seeing a mix of field stars and true cluster members should be borne in mind when viewing an open cluster through the eyepiece.

M 40

Constellation	Object type	RA, Dec	Approx. transit date at local midnight	Distance
Ursa Major	Double star	12h 22.4m, +58° 05′	April 21	5 hundred light years
Age	**Apparent size**	**Magnitude**	**Sky Atlas 2000.0 chart**	**Herald–Bobroff chart**
		8	2	D35

This is simply a double star (mag. 9.0 and 9.3, separation 50″, position angle 83° i.e. the two stars lie along a nearly E–W line), also known as Winnecke 4. Messier included it when looking for a nebula reported by Hevelius in this region.

M 41 (NGC 2287)

Constellation	Object type	RA, Dec	Approx. transit date at local midnight	Distance
Canis Major	Open cluster	06h 46.0m, −20° 45′	January 11	2 thousand light years
Age	**Apparent size**	**Magnitude**	**Sky Atlas 2000.0 chart**	**Herald–Bobroff chart**
2 hundred million years	38′	4.5	19	C70/D22

It has a diameter of about 20 light years. A large percentage (perhaps as high as 80%) of its stars are binary stars. Open clusters are beasts of the Galactic disk, and are stripped of their stars over time by gravitational interaction with material in the disk as they jostle about the disk while rotating with it. For an open cluster in our vicinity of the Galaxy, like this one, typical lifetimes are thought to be a little over 1/2 billion years, so that this cluster is approaching middle age. The cluster has about 70 members with magnitude brighter than 12, although our view of the cluster to this magnitude is "contaminated" by about as many field stars as cluster members (see M 39).

M 42 (NGC 1976)

Constellation	Object type	RA, Dec	Approx. transit date at local midnight	Distance
Orion	Emission and reflection nebula	05h 35.3m, −05° 23′	December 24	1.5 thousand light years
Age	Apparent size	Magnitude	Sky Atlas 2000.0 chart	Herald–Bobroff chart
	1.5° × 1°	4.0	11	C53/D24

Nicknamed the "Orion Nebula". This is the apparently brightest and one of the closest ionized hydrogen star-forming regions (or HII regions) in our sky. It is lit up by the stars in a very young cluster (only a few hundred thousand years old) situated in the heart of the nebula. However, many of the stars in this cluster are obscured by material in the nebula. Light from these stars is both scattered off dust, particularly in the outer regions of the nebula, and re-emitted by gas, particularly in the inner regions, making this both a reflection and emission nebula. Four stars in the cluster form a quadrangle (with sides of about 10–20″) called the "Trapezium" and are all part of the multiple star system θ^1 Orionis. Professional telescopic studies indicate θ^1 Orionis contains 14 stars, only six of which can be seen, including the four in the Trapezium, under good seeing conditions in moderate amateur telescopes. θ^1 Orionis is in fact a wide double star with θ^2 Orionis (itself a triple star, so that θ Orionis consists of 17 stars!). The brightest (most southern) star in the Trapezium quadrangle (θ^1 Orionis C) is responsible for the ionization of the nebula. M 42 is actually only a small "blister" on the near side of a much larger cloud of gas and dust (the Orion A complex). The Orion A complex is itself part of an even larger group of giant molecular gas clouds (the Orion–Monoceros complex) that extend 30° in a SE–NW direction and sit about 5 hundred light years below the Galactic central plane. The Orion–Monoceros complex may have formed as the result of the infall of a gargantuan cloud onto the Galactic disk from below. The 1.5° apparent dimension of M 42 corresponds to a diameter of 40 light years.

M 43 (NGC 1982)

Constellation	Object type	RA, Dec	Approx. transit date at local midnight	Distance
Orion	Emission nebula	05h 35.5m, −05° 16′	December 24	1.5 thousand light years
Age	Apparent size	Magnitude	Sky Atlas 2000.0 chart	Herald–Bobroff chart
	20′	9.0	11	C53/D24

This is part of the same gas and dust cloud as M 42 (the Orion A complex), lying just NE of M 42, separated from M 42 by a dust lane between the two. M 43 has relatively little dust, so its light is largely from gas emission (it is an ionized hydrogen, i.e. HII, region). The nebula is visible because it is ionized by the bright variable star NU Orionis (i.e. HD 37061, mag. 7) in its center. Professional telescopic studies find planet-forming disks ("proplyds") are present around several stars in M 43 (as well as in M 42).

M 44 (NGC 2632)

Constellation	Object type	RA, Dec	Approx. transit date at local midnight	Distance
Cancer	Open cluster	08h 40.0m, +19° 40′	February 9	6 hundred light years

Age	Apparent size	Magnitude	Sky Atlas 2000.0 chart	Herald–Bobroff chart
7 hundred million years	1.6°	3.1	6	C32

Nicknamed the "Beehive Cluster" or "Praesepe" (which means manger in Latin, its common-use anglicized pronunciation being pree-SEE-pee). Recent work suggests that M 44 is actually two merging clusters. The total cluster mass is about 7 hundred solar masses. M 44 and the Hyades (60° W) may be part of a single, moving group (some suggest the two formed from a single gaseous cloud). M 44's size of 1.6° corresponds to a diameter of about 16 light years.

M 45

Constellation	Object type	RA, Dec	Approx. transit date at local midnight	Distance
Taurus	Open cluster	03h 47.0m, +24° 07′	November 27	4 hundred light years

Age	Apparent size	Magnitude	Sky Atlas 2000.0 chart	Herald–Bobroff chart
1 hundred million years	1.8°	1.2	4	C36/D26

Nicknamed the "Pleiades" (pronounced PLEE-ah-deez) or the "Seven Sisters" (although only six stars are visible with the naked eye in light polluted skies). The cluster has a mass of about 8 hundred Suns, including an inexplicably large number of white dwarf stars for its young age. The Orion Nebula cluster (see M 42) is believed to be a close replica of what M 45 was like when it was a few hundred thousand years old. It is thought that about 70% of the stars in M 45 are binaries (roughly the same as the fraction of binary stars present elsewhere in our Galaxy).

M 46 (NGC 2437)

Constellation	Object type	RA, Dec	Approx. transit date at local midnight	Distance
Puppis	Open cluster	07h 41.8m, −14° 49′	January 25	5 thousand light years

Age	Apparent size	Magnitude	Sky Atlas 2000.0 chart	Herald–Bobroff chart
3 hundred million years	27′	6.1	12	C51/D22

Its 27′ size corresponds to a diameter of about 40 light years. It lies about 4 hundred light years above the Galactic central plane at a distance of about 35 thousand light years from the Galactic center. Although the planetary nebula NGC 2438 lies on the NE edge of M 46 (see NGC 2438), their differing relative velocities and the young age of M 46 together suggest that NGC 2438 is not part of M 46. Whether NGC 2438 is a foreground or background object remains uncertain (it has been suggested it is a foreground object lying about 2 thousand light years closer than M 46).

M 47 (NGC 2422)

Constellation	Object type	RA, Dec	Approx. transit date at local midnight	Distance
Puppis	Open cluster	07h 36.6m, −14° 29′	January 24	2 thousand light years
Age	**Apparent size**	**Magnitude**	**Sky Atlas 2000.0 chart**	**Herald–Bobroff chart**
1 hundred million years	30′	4.4	12	C51/D22

Its diameter is about 18 light years. Interstellar dust between us and the stars in this cluster cause its stars to appear dimmer by only a few tenths of a magnitude, which is much less than the average two magnitudes of dimming for every kiloparsec (3.26 thousand light years) that is typical when light travels in the central plane of our Galaxy. Several Be stars are known in this cluster, the brightest of which is HD 60856 at mag. 8 and is readily visible in amateur telescopes (see Figure 2422). Be stars are B-type stars that are peculiar because of hydrogen Balmer emission lines in their spectra, due to atomic transitions in material expelled by high rotational velocities into a circumstellar disk in the equatorial plane of the star.

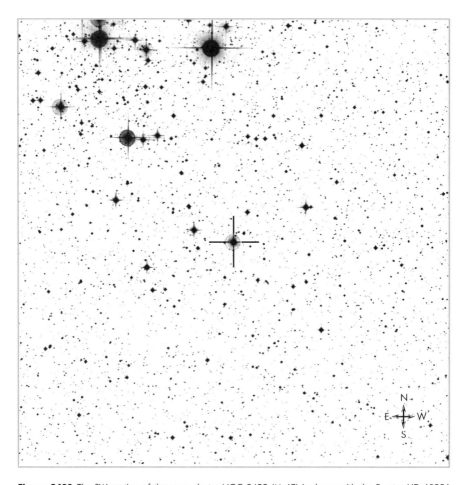

Figure 2422 The SW portion of the open cluster NGC 2422 (M 47) is shown with the Be star HD 60856 in the center of the figure (marked by cross-hairs, not diffraction spikes like the other stars not in the center of the figure). The area shown is 30′ × 30′. From the Digitized Sky Survey (Space Telescope Science Institute) based on photographic data of the National Geographic Society – Palomar Observatory Sky Survey (POSS I).

M 48 (NGC 2548)

Constellation	Object type	RA, Dec	Approx. transit date at local midnight	Distance
Hydra	Open cluster	08h 13.7m, –05° 45′	February 2	2 thousand light years
Age	**Apparent size**	**Magnitude**	**Sky Atlas 2000.0 chart**	**Herald–Bobroff chart**
3 hundred million years	54′	5.8	12	C51

Professional telescopic studies find 165 stars with membership probability > 70%. It lies about 29 thousand light years from the center of our Galaxy and a little more than 5 hundred light years above the Galactic central plane. Its current position is close to its maximum excursion of 6 hundred light years from this plane, having crossed it eight times in its lifetime, while making slightly more than one revolution about the Galactic center. It has a diameter of about 30 light years. Messier's discovery and subsequent listing of this object resulted in an error in its quoted position, so that its location in some old star charts (before T.F. Morris' correction of this error in 1959) is incorrect.

M 49 (NGC 4472)

Constellation	Object type	RA, Dec	Approx. transit date at local midnight	Distance
Virgo	Elliptical galaxy	12h 29.8 m, +08° 00′	April 23	55 million light years
Age	**Apparent size**	**Magnitude**	**Sky Atlas 2000.0 chart**	**Herald–Bobroff chart**
	9.8′ × 8.2′	8.4	14	C48/D34/E11

This is a giant elliptical galaxy and the brightest member of the Virgo galaxy cluster, which is the nearest galaxy cluster to us. Galaxy clusters are the largest stable structures in the Universe, although only about 1% of galaxies belong to galaxy clusters. The Virgo galaxy cluster contains many hundreds of galaxies, which are bunched into several subclusters with M 87 and M 49 being principal members of the two main subclusters. M 49 lies perhaps a few million light years closer than M 87. Occupying a roughly rectangular shape in the sky (8° E–W × 16° N–S), the major concentration of the Virgo cluster extends several tens of millions of light years around the cluster center near M 87 (but, for simplicity, all its members are listed herein as being at the same distance). M 49 is thought to be falling toward the center of the Virgo cluster at about a thousand km/s. Much of the mass in the Virgo cluster is dark matter, with the total mass of the Virgo cluster being many hundreds of trillions of solar masses. M 49 contains about 6 thousand globular clusters of two different ages, which suggests M 49 may be the product of a merger event in its past. The center of M 49 is thought to contain a supermassive black hole with a mass of about 1/2 billion Suns. The luminous mass of M 49 is roughly 2 hundred billion Suns, but its total mass (including dark matter in its halo) may be almost 2 trillion Suns. Its optical diameter is about 150 thousand light years.

M 50 (NGC 2323)

Constellation	Object type	RA, Dec	Approx. transit date at local midnight	Distance
Monoceros	Open cluster	07h 02.1m, −08° 23′	January 15	3 thousand light years
Age	**Apparent size**	**Magnitude**	**Sky Atlas 2000.0 chart**	**Herald–Bobroff chart**
1 hundred million years	16′	5.9	12	C52/D22

It has roughly 2 hundred members, a diameter of about 15 light years and lies about 70 light years below the central Galactic plane. The interstellar gas and dust in the disk of our Galaxy is not uniformly distributed at light year length scales, but instead is clumped into patches with typical masses of a few hundred Suns. Open clusters are thought to form from such interstellar clouds of gas and dust.

M 51 (NGC 5194)

Constellation	Object type	RA, Dec	Approx. transit date at local midnight	Distance
Canes Venatici	Spiral galaxy	13h 29.9m, +47° 12′	May 8	30 million light years
Age	**Apparent size**	**Magnitude**	**Sky Atlas 2000.0 chart**	**Herald–Bobroff chart**
	10.8′ × 6.6′	8.4	7	C29

Nicknamed the "Whirlpool Galaxy". It is a grand-design spiral, meaning that it has two symmetrically placed spiral arms that extend over most its visible disk in professional telescopic images. What appears as a bridge connecting M 51 to its nearby companion galaxy NGC 5195 (4′ NNE – see NGC 5195) is actually an optical illusion – a spiral arm of M 51 is superimposed on NGC 5195 with NGC 5195 actually lying on the far side of M 51 (by perhaps 1/2 million light years), although the two galaxies have had one or more close encounters in the past (see NGC 5195). Gravitational interaction with NGC 5195 is thought to be triggering star formation in M 51, with about four new stars per year forming in M 51. This is similar to the number forming per year in the Milky Way, but note that our Galaxy has a mass about 10 times that of M 51. More than a thousand ionized hydrogen (HII) star-forming regions have been identified in M 51 by the Hubble Space Telescope, with diameters averaging about a hundred light years and having masses up to about a thousand Suns. A number of these HII regions can be seen as bright knots in amateur telescopes. M 51 is a LINER galaxy (see M 81/NGC 3031 for explanation) and has an optical diameter of almost 1 hundred thousand light years.

M 52 (NGC 7654)

Constellation	Object type	RA, Dec	Approx. transit date at local midnight	Distance
Cassiopeia	Open cluster	23h 24.8m, +61° 36′	October 6	4 thousand light years
Age	**Apparent size**	**Magnitude**	**Sky Atlas 2000.0 chart**	**Herald–Bobroff chart**
1 hundred million years	13′	6.9	3	C03/D06

This is a relatively rich cluster. To magnitude 15.0, there is nearly one star for every square arcsecond of sky in its densest parts (although about 1 in 10 of these is a field star and not a cluster member – see M 39). To magnitude 14.5, a total of about 130 stars belong to the cluster (with only about 30 field stars "contaminating" the cluster), with these cluster members having masses about 2–5 times that of our Sun. To magnitude 19.5, over 6 thousand stars belong to the cluster (with about the same number of field stars present), with most of these dimmer cluster members having masses near that of our Sun. The stars in this cluster appear to have a much larger spread of ages (tens of millions of years) than most open clusters (where the stars are typically only a few million years apart in age). Gas and dust between us and the cluster dim the stars in this cluster considerably (by a few magnitudes).

M 53 (NGC 5024)

Constellation	Object type	RA, Dec	Approx. transit date at local midnight	Distance
Coma Berenices	Globular cluster	13h 12.9m, +18° 10′	May 4	60 thousand light years
Age	**Apparent size**	**Magnitude**	**Sky Atlas 2000.0 chart**	**Herald–Bobroff chart**
12–14 billion years	14.4′	7.6	14	C29

Its mass is about 3/4 million Suns (which is about a millionth the mass of our entire Galaxy). It lies nearly directly "below" us from the Galactic central plane in the halo of our Galaxy (see M 2 for the meaning of "halo"). Its orbit keeps it out in the halo, with a period of close to a billion years. It follows a path that takes it well over a hundred thousand light years from the Galactic center and never closer than about 50 thousand light years to the Galactic center. Its orbit is highly inclined (by about 60°) to the disk of our Galaxy. M 53's apparent size corresponds to a diameter of about 250 light years. The nearby globular cluster NGC 5053 (1° ESE) lies nearly directly "above" M 53 in three-dimensional space, being a little less than 10 thousand light years closer to us (in a direction nearly perpendicular to the disk of our Galaxy).

M 54 (NGC 6715)

Constellation	Object type	RA, Dec	Approx. transit date at local midnight	Distance
Sagittarius	Globular cluster	18h 55.1m, −30° 29′	July 30	80 thousand light years
Age	Apparent size	Magnitude	Sky Atlas 2000.0 chart	Herald–Bobroff chart
12–14 billion years	9.1′	7.6	22	C61

This globular cluster is not part of our Galaxy, but instead belongs to a nearby satellite galaxy (that goes by the cumbersome name "Sagittarius Dwarf Elliptical Galaxy"). It has been suggested that M 54 may actually be its nucleus. This companion galaxy is in the process of being gravitationally disrupted by our Galaxy. Models predict that it, along with M 54, will collide with the disk of our Galaxy in several tens of millions of years, having had a past such collision about 2 hundred million years ago. M 54 is one of the most massive known globular clusters with a mass of about 1.5 million Suns (which is about 1/40 the mass of the galaxy it belongs to). Its size of 7.6′ corresponds to a diameter of about 2 hundred light years.

M 55 (NGC 6809)

Constellation	Object type	RA, Dec	Approx. transit date at local midnight	Distance
Sagittarius	Globular cluster	19h 40.0m, −30° 58′	August 10	20 thousand light years
Age	Apparent size	Magnitude	Sky Atlas 2000.0 chart	Herald–Bobroff chart
12–14 billion years	19.1′	6.3	22	C60

It has a mass of about 1/4 million Suns and a diameter of about a hundred light years. It takes a little over a hundred million years or so to complete an orbit about our Galaxy, always staying within about 20 thousand light years of the Galactic center but never swinging closer than about 5 thousand light years from the Galactic center, all in a path that is highly inclined to the disk of our Galaxy (by about 50–60°). It contains 74 known "blue stragglers" (see NGC 6633), which are stars that are paradoxically far more blue and luminous than expected for reasons that remain unknown (but may involve the coalescence of stars).

M 56 (NGC 6779)

Constellation	Object type	RA, Dec	Approx. transit date at local midnight	Distance
Lyra	Globular cluster	19h 16.6m, +30° 11′	August 4	32 thousand light years
Age	Apparent size	Magnitude	Sky Atlas 2000.0 chart	Herald–Bobroff chart
12–14 billion years	5.0′	8.3	8	C24/D09

It has a mass of about 2 hundred thousand Suns and a diameter of about 50 light years. Although its orbit is roughly circular and lies nearly in the Galactic disk (inclined by only about 15° to the central plane), like almost all globular clusters in our Galaxy, it does not orbit with the disk material at a constant radius from the Galactic center. Instead its path is thought to take it out as far as about 40 thousand light years away from the Galactic center and within a few thousand light years of the Galactic center, although it takes about the same amount of time to complete an orbit as our Sun (i.e. 1/4 billion years).

M 57 (NGC 6720)

Constellation	Object type	RA, Dec	Approx. transit date at local midnight	Distance
Lyra	Planetary nebula	18h 53.6m, +33° 02′	July 29	2 thousand light years
Age	Apparent size	Magnitude	Sky Atlas 2000.0 chart	Herald–Bobroff chart
	1.2′	9	8	C25

Nicknamed the "Ring Nebula". This planetary nebula is thought to have been created about a thousand years ago when an old star blew off its outer layers. In professional telescopes the nebula extends out to nearly 4′ in diameter (with its outer halo 5 thousand times dimmer in surface brightness than the ring). Its three-dimensional structure remains inconclusively understood, but its main part may consist of an ellipsoidal shell (like the skin of an air-ship/dirigible as in the Hindenberg or the Good Year blimp) that we are looking at nearly end-on. This shell is thought to be more dense at its mid-section (half-way along the "dirigible") so that the ring we see is merely a torus of denser material at the mid-section of the ellipsoidal shell. The denseness of the ring is thought to be a relic of the preferential ejection of mass by the central star in its equatorial plane (in a "superwind" – see M 27). The shell is expanding outward at about 50 km/s. Invisible (UV) radiation from the hot central star ionizes the atoms in the shell, and electrons recombining with these ionized atoms cause optical photons to be emitted that make the nebula visible to us.

M 58 (NGC 4579)

Constellation	Object type	RA, Dec	Approx. transit date at local midnight	Distance
Virgo	Barred spiral galaxy	12h 37.7m, +11° 49′	April 25	55 million light years
Age	Apparent size	Magnitude	Sky Atlas 2000.0 chart	Herald–Bobroff chart
	6.0′ × 4.8′	9.7	14	C47/D33/E09

It is part of the Virgo galaxy cluster (see M 49/NGC 4472). M 58 has a mass of about 3 hundred million Suns. It is a Seyfert galaxy (or possibly a LINER – see M 81) in which emission from the nucleus is thought to occur due to accretion of matter onto a supermassive central black hole. Its optical diameter is about 1 hundred thousand light years.

M 59 (NGC 4621)

Constellation	Object type	RA, Dec	Approx. transit date at local midnight	Distance
Virgo	Elliptical galaxy	12h 42.0m, +11° 39′	April 26	55 million light years
Age	Apparent size	Magnitude	Sky Atlas 2000.0 chart	Herald–Bobroff chart
	5.3′ × 4.0′	9.6	14	C47/D33/E09

It is part of the Virgo galaxy cluster (see M 49/NGC 4472). The inner core of this galaxy is unusual because it is thought to be counter-rotating from the rest of the galaxy as well as having a different chemical makeup from, and being younger than, the rest of the galaxy, perhaps due to a past accretion event. This inner core has a diameter of several hundred light years and apparent size of about an arcsecond. Professional telescopes find M 59 has a disk that emits about 16% of the light from this galaxy. M 59 also has a circumnuclear disk (diameter of 5–7″) in professional telescopic studies. The center of this galaxy is believed to contain a supermassive black hole (with a mass of about 3 hundred million Suns). M 59 contains about 2 thousand globular clusters.

M 60 (NGC 4649)

Constellation	Object type	RA, Dec	Approx. transit date at local midnight	Distance
Virgo	Elliptical galaxy	12h 43.7m, +11° 33′	April 26	55 million light years
Age	Apparent size	Magnitude	Sky Atlas 2000.0 chart	Herald–Bobroff chart
	7.6′ × 6.2′	8.8	14	C47/D33/E09

It is part of the Virgo galaxy cluster (see M 49/NGC 4472). M 60 is thought to contain a supermassive central black hole (with a mass of perhaps a billion Suns). Its nearby companion galaxy NGC 4647 (barred spiral galaxy, 2.5′ NW, mag. 11.3) may lie somewhat in front of M 60, but this remains uncertain. The distance between the centers of these two in the sky corresponds to an actual distance of about 40 thousand light years. M 60 contains about 5 thousand globular clusters (about 30 times as many as in our Galaxy).

M 61 (NGC 4303)

Constellation	Object type	RA, Dec	Approx. transit date at local midnight	Distance
Virgo	Barred spiral galaxy	12h 21.9 m, +04° 28'	April 21	55 million light years
Age	**Apparent size**	**Magnitude**	**Sky Atlas 2000.0 chart**	**Herald–Bobroff chart**
	5.5' × 2.9'	9.6	14	C48/D34/E12

This is one of fewer than about 10 galaxies that have had four or more recorded super-novae. It has a total mass of about 70 billion Suns, a diameter of about 90 thousand light years and is thought to be a Seyfert galaxy (see NGC 3227). It is part of the Virgo cluster (see M 49/NGC 4472) and is a grand-design spiral, meaning that it has two symmetrically placed spiral arms that extend over most its visible disk in professional telescopic images. Many giant, ionized hydrogen (star-forming) regions are present along its spiral arms, giving the arms an uneven brightness along their length in large amateur telescopes. It also has a massive circumnuclear star-forming disk (with a diameter of about 10″ and a mass of about 50 million Suns) that itself has a bar and spiral structure. Total star-formation rates in this galaxy are probably 1–2 solar masses/year (about half that occurring in our Galaxy, but our Galaxy is about 10 times as massive). It is probably interacting with nearby NGC 4303A (10' NE, mag. 13, listed as NGC 4303 on Herald–Bobroff chart E12) and NGC 4292 (12' NW, mag. 12.2).

M 62 (NGC 6266)

Constellation	Object type	RA, Dec	Approx. transit date at local midnight	Distance
Ophiuchus	Globular cluster	17h 01.2m, –30° 07'	July 1	20 thousand light years
Age	**Apparent size**	**Magnitude**	**Sky Atlas 2000.0 chart**	**Herald–Bobroff chart**
12–14 billion years	14.1'	6.5	22	C62/D13

It has a mass of almost a million Suns. Its size of 14.1' corresponds to a diameter of about 80 light years. It lies on the edge of the Galactic disk, in the direction of the Galactic center from us and is a "bulge cluster" (meaning it spends its time orbiting around the central, spherical, 15 thousand light year diameter bulge of our Galaxy). It is one of fewer than 20 globular clusters known to contain a pulsar (with a pulse period of 5.24 milliseconds; see M 1 for the meaning of "pulsar"). The pulsar in this case is a binary star consisting of a neutron star and a white dwarf (orbiting the neutron star every 3.8 days). The core of M 62 is thought to be "collapsed" (see NGC 6284 for explanation).

M 63 (NGC 5055)

Constellation	Object type	RA, Dec	Approx. transit date at local midnight	Distance
Canes Venatici	Spiral galaxy	13h 15.8 m, +42° 02′	May 4	30 million light years
Age	**Apparent size**	**Magnitude**	**Sky Atlas 2000.0 chart**	**Herald– Bobroff chart**
	12.6′ × 7.5′	8.6	7	C29/D30

Nicknamed the "Sunflower Galaxy". It is part of a gravitationally bound group of four galaxies that includes M 51 (NGC 5194) and NGC 5023 (spiral galaxy, mag. 12.3, 2° NNW), in addition to the dimmer UGC 8320 (irregular galaxy, mag. 13, 4° N). M 63 is a so-called "flocculent" spiral galaxy, meaning that it lacks any obvious azimuthally symmetric spiral arm pattern (see NGC 3521 for further explanation). It has a mass of about 140 billion Suns and is a "low-ionization nuclear emission region" galaxy (or "LINER" – see M 81/NGC 3031 for explanation). It has an optical diameter of a little over 1 hundred thousand light years. Beyond its optically visible regions the galaxy consists of a neutral hydrogen (HI) disk (not visible in amateur telescopes) that is "warped" so that the far outer edge of its HI disk lies in a plane that is skewed by perhaps about 20° from the rest of the galaxy. Warps occur in the HI distribution of about half of all galactic disks.

M 64 (NGC 4826)

Constellation	Object type	RA, Dec	Approx. transit date at local midnight	Distance
Coma Berenices	Spiral galaxy	12h 56.7 m, +21° 41′	April 30	16 million light years
Age	**Apparent size**	**Magnitude**	**Sky Atlas 2000.0 chart**	**Herald– Bobroff chart**
	10.3′ × 5.0′	8.5	7	C29

Nicknamed the "Black-Eye Galaxy" (because of a dark arc-shaped dust region on its NE side, which is a challenge to discern in amateur telescopes). It has an optical diameter of about 50 thousand light years. The nucleus of this galaxy is chemically different from the rest of the galaxy (and is said to be "chemically decoupled"). In addition, the gas in the outer disk (radii > about 1′ and containing a hundred million solar masses) counter-rotates from rest of the galaxy, including the stars (which all rotate the same way). This highly unusual situation may have its origin in the past accretion of a gas-rich dwarf satellite galaxy. The dust that gives the galaxy its "black eye" rotates with the stars. M 64 is thought to be a LINER galaxy driven by a starburst in its nucleus (see M 81/NGC 3031 for explanation).

M 65 (NGC 3623)

Constellation	Object type	RA, Dec	Approx. transit date at local midnight	Distance
Leo	Barred spiral galaxy	11h 18.9 m, +13° 05′	March 21	35 million light years
Age	**Apparent size**	**Magnitude**	**Sky Atlas 2000.0 chart**	**Herald–Bobroff chart**
	9.0′ × 2.3′	9.3	13	C49/D36

M 65 is nearly edge-on with its rotation axis inclined by 74° from our line-of-sight. It has an optical diameter of about 90 thousand light years. It is part of a gravitationally bound group of galaxies that includes NGC 3593 (1° SW), as well as nearby M 66 (see M 66/NGC 3627) and NGC 3628 (see NGC 3628) with which it forms the "Leo Triplet". Unlike M 66 and NGC 3628 (which have strongly interacted in the past – see NGC 3628), M 65 does not seem too much affected by the presence of the other galaxies in this group. M 65 is a "low-ionization nuclear emission region" (LINER) galaxy (see M 81/NGC 3031 for explanation).

M 66 (NGC 3627)

Constellation	Object type	RA, Dec	Approx. transit date at local midnight	Distance
Leo	Barred spiral galaxy	11h 20.3 m, +12° 59′	March 21	35 million light years
Age	**Apparent size**	**Magnitude**	**Sky Atlas 2000.0 chart**	**Herald–Bobroff chart**
	9.1′ × 4.1′	8.9.	13	C49/D36

M 66 is part of the "Leo Triplet" that includes nearby M 65 (see M 65/NGC 3623) and NGC 3628, which form a gravitationally bound group of four galaxies (the fourth member being nearby NGC 3593). Professional telescopes find a quarter million light year (40′) long plume of stars and gas extending to the east of NGC 3628, with a mass of hundreds of millions of Suns, that is thought to be the result of an interaction with M 66 nearly a billion years ago. M 66 is one of fewer than 20 galaxies that have had three or more recorded supernovae. It has an optical diameter of about 90 thousand light years.

M 67 (NGC 2682)

Constellation	Object type	RA, Dec	Approx. transit date at local midnight	Distance
Cancer	Open cluster	08h 50.8m, +11° 49′	February 11	3 thousand light years
Age	**Apparent size**	**Magnitude**	**Sky Atlas 2000.0 chart**	**Herald–Bobroff chart**
4 billion years	30′	6.9	12	C50

This is one of the oldest open clusters known. Open clusters that orbit closer in to the Galactic center are disrupted more readily, so that almost all old open clusters have orbits that stay at least as far from the Galactic center as we are located (i.e. at radii > 25 thousand light years or so). Professional telescopes find about 5 hundred members (down to magnitude 17) in this cluster. Its size of 30′ corresponds to a diameter of about 25 light years.

M 68 (NGC 4590)

Constellation	Object type	RA, Dec	Approx. transit date at local midnight	Distance
Hydra	Globular cluster	12h 39.5m, −26° 45′	April 26	30 thousand light years
Age	**Apparent size**	**Magnitude**	**Sky Atlas 2000.0 chart**	**Herald– Bobroff chart**
12–14 billion years	9.8′	7.8	21	C66

It has a mass of about 3 hundred thousand Suns. Its size of 9.8′ corresponds to a diameter of almost one hundred light years. Like most globular clusters, it does not orbit our Galaxy with the Galactic disk as we do. Instead it follows an orbit that takes it out as far as about 1 hundred thousand light years from the Galactic center and then back in as close as about 30 thousand light years on an elliptical path (with eccentricity 0.5) inclined to the Galactic disk (by about 30°), taking about 1/2 billion years to make one revolution around our Galaxy. It is very "metal-poor" (meaning it is sparse in elements heavier than helium).

M 69 (NGC 6637)

Constellation	Object type	RA, Dec	Approx. transit date at local midnight	Distance
Sagittarius	Globular cluster	18h 31.4m, −32° 21′	July 24	37 thousand light years
Age	**Apparent size**	**Magnitude**	**Sky Atlas 2000.0 chart**	**Herald– Bobroff chart**
12–14 billion years	7.1′	7.6	22	C61

It lies almost directly below the Galactic center (about 5 thousand light years below the Galactic central plane), about 10 thousand light years on the other side of the Galactic center from us. It has a mass of about 3 hundred thousand Suns and its size corresponds to a diameter of about 75 light years. It is thought to be a "bulge cluster" (meaning it spends its time orbiting around the central, spherical, 15 thousand light year diameter bulge of our Galaxy). It is a "metal-rich" cluster (meaning it contains significant amounts of elements heavier than helium). About 1/4 of the globular clusters in our Galaxy are considered metal-rich.

M 70 (NGC 6681)

Constellation	Object type	RA, Dec	Approx. transit date at local midnight	Distance
Sagittarius	Globular cluster	18h 43.2m, −32° 18′	July 27	30 thousand light years
Age	**Apparent size**	**Magnitude**	**Sky Atlas 2000.0 chart**	**Herald– Bobroff chart**
12–14 billion years	7.8′	7.9	22	C61

It has a mass of about 2 hundred thousand Suns and a diameter of about 70 light years. Like its neighbor M 69, this is thought to be a "bulge cluster" (meaning it spends its time orbiting around the bulge of our Galaxy – see M 69/NGC 6637). Its core is thought to have "collapsed", the result of an instability that causes the stars in its core to confine themselves to an unusually small region (see NGC 6284).

M 71 (NGC 6838)

Constellation	Object type	RA, Dec	Approx. transit date at local midnight	Distance
Sagitta	Globular cluster	19h 53.8m, +18° 47′	August 14	12 thousand light years

Age	Apparent size	Magnitude	Sky Atlas 2000.0 chart	Herald–Bobroff chart
12–14 billion years	6.1′	8.2	16	C24/D09

This is one of the closest globular clusters to us. With a mass of less than about 40 thousand Suns and a diameter of a little over 20 light years, this is a sparse globular cluster. It is one of only a few globular clusters whose orbit stays in the disk of our Galaxy (so it is said to be a "disk cluster"). Its orbit is highly elliptical (with a minor to major axis ratio of 0.2), and it takes about 160 million years to complete one orbit around our Galaxy. This cluster contains significant amounts of elements heavier than helium (and so is said to be "metal-rich" – only about 1/4 of the globular clusters in our Galaxy are metal-rich, the rest being metal-poor).

M 72 (NGC 6981)

Constellation	Object type	RA, Dec	Approx. transit date at local midnight	Distance
Aquarius	Globular cluster	20h 53.5m, −12° 32′	August 30	65 thousand light years

Age	Apparent size	Magnitude	Sky Atlas 2000.0 chart	Herald–Bobroff chart
12–14 billion years	5.1′	9.3	16	C41

It has a mass of about 2 hundred thousand Suns and a diameter of about one hundred light years. It lies in the halo of our Galaxy (see M 2 for the meaning of "halo"). Like 3/4 of the globular clusters in our Galaxy, this is a "metal-poor" cluster (meaning it has a low abundance of elements heavier than helium). It rotates about our Galaxy in a retrograde direction (i.e. opposite to the Sun's motion around the Galaxy), which has led to the suggestion that it was adopted in a merger with another galaxy, but this is not certain.

M 73 (NGC 6994)

Constellation	Object type	RA, Dec	Approx. transit date at local midnight	Distance
Aquarius	Asterism	20h 58.9m, −12° 38′	August 31	

Age	Apparent size	Magnitude	Sky Atlas 2000.0 chart	Herald–Bobroff chart
	2.8′	9	16	C41

Although there has been some confusion as to whether this is an open cluster or not, recent data shows that the four stars at this location are not a cluster but are simply an asterism (i.e. a pattern of physically unrelated stars on the sky).

M 74 (NGC 628)

Constellation	Object type	RA, Dec	Approx. transit date at local midnight	Distance
Pisces	Spiral galaxy	01h 36.7 m, +15° 47′	October 25 (Standard Time)	30 million light years
Age	Apparent size	Magnitude	Sky Atlas 2000.0 chart	Herald–Bobroff chart
	10.0′ × 9.4′	9.4	10	C38

It has a mass of about 330 billion Suns and an optical diameter of almost 1 hundred thousand light years. It is nearly face-on (its inclination angle, which is the angle between its axis of rotation and our line-of-sight, is about 7°). Hundreds of ionized hydrogen (HII) star-forming regions (like M 42) have been identified in this galaxy in professional telescopes, some of which can be seen as bright knots in large amateur telescopes. All told, these star-forming regions produce about one new star per year. M 74 is part of a gravitationally bound group of seven galaxies that includes NGC 660 (2° 38′ SE) as well as five dimmer galaxies (the brightest of which, at mag. 13, are UGC 1195, 22′ NNW of NGC 660 and UGC 1200, 29′ S of NGC 660). Professional telescopic studies show that beyond its optically visible disk is an extended disk of atomic hydrogen reaching out more than twice its optical diameter. This extended disk is distorted for reasons that remain unknown, but may involve past interactions with other galaxies.

M 75 (NGC 6864)

Constellation	Object type	RA, Dec	Approx. transit date at local midnight	Distance
Sagittarius	Globular cluster	20h 06.1m, −21° 55′	August 17	60 thousand light years
Age	Apparent size	Magnitude	Sky Atlas 2000.0 chart	Herald–Bobroff chart
12–14 billion years	4.6′	8.5	23	C60

It has a mass of about 1/2 million Suns and a diameter of about 80 light years. Like 3/4 of the globular clusters in our Galaxy, it is "metal-poor" (i.e. it has a low abundance of "metals", which in astrophysics means elements heavier than helium). It lies on the other side of the Galaxy from us, well below the Galactic central plane (by about 30 thousand light years).

M 76 (NGC 650/651)

Constellation	Object type	RA, Dec	Approx. transit date at local midnight	Distance
Perseus	Planetary nebula	01h 42.3m, +51° 34′	October 27 (Standard Time)	4 thousand light years
Age	Apparent size	Magnitude	Sky Atlas 2000.0 chart	Herald–Bobroff chart
	2′	11	1	C20

Nicknamed the "Little Dumbbell Nebula". Although different parts are expanding at different rates, typical expansion velocities are about 50 km/s. It has a diameter of about 2 light years. The two lobes of the "bowtie" shape are each three-dimensional expanding bubbles, inclined at 75° from our line-of-sight with the NW lobe pointing toward us.

M 77 (NGC 1068)

Constellation	Object type	RA, Dec	Approx. transit date at local midnight	Distance
Cetus	Spiral galaxy	02h 42.7m, –00° 01′	November 11	50 million light years
Age	**Apparent size**	**Magnitude**	**Sky Atlas 2000.0 chart**	**Herald–Bobroff chart**
	7.3′ × 6.3′	8.9	10	C55

This is a well-known Seyfert galaxy, in which emission from the nucleus is thought to occur due to accretion of matter onto a massive central black hole (which is thought to have a mass of about 20 million Suns in this galaxy). It is part of a gravitationally bound group of six galaxies that includes NGC 1055 (30′ NNW), NGC 1073 (1° 25′ N), as well as the much dimmer, mag. 13, UGC 2275 and UGC 2302. M 77 also contains water-vapor "masers" in its central region. Maser stands for "microwave amplification by stimulated emission of radiation", the physics of which is the microwave equivalent of a laser, except that lasers are usually designed to produce a beam, while astronomical masers yield emission that radiates from a roughly spherical region. M 77 has an optical diameter of about 110 thousand light years.

M 78 (NGC 2068)

Constellation	Object type	RA, Dec	Approx. transit date at local midnight	Distance
Orion	Reflection nebula	05h 46.8m, +00° 05′	November 11	1.5 thousand light years
Age	**Apparent size**	**Magnitude**	**Sky Atlas 2000.0 chart**	**Herald–Bobroff chart**
	8′	8	11	C53/D24

M 78 is a star-forming region. It is illuminated by the double star HD 290862, which is the more southern of the two bright stars seen in this nebula in amateur telescopes. M 78 is part of a giant molecular cloud (the Orion B cloud) that has a size of about 8° in professional telescopic studies (and also includes the fellow star-forming regions NGC 2071, NGC 2023 and NGC 2024) and is part of the much larger Orion–Monoceros complex (see M 42). Its size of 8′ corresponds to a diameter of a few light years.

M 79 (NGC 1904)

Constellation	Object type	RA, Dec	Approx. transit date at local midnight	Distance
Lepus	Globular cluster	05h 24.2m, –24° 31′	December 22	40 thousand light years
Age	**Apparent size**	**Magnitude**	**Sky Atlas 2000.0 chart**	**Herald–Bobroff chart**
12–14 billion years	7.8′	7.7	19	C71

It has a mass of almost 4 hundred thousand Suns and a diameter of about one hundred light years. Like most globular clusters it does not orbit our Galaxy with the Galactic disk as we do. Instead it follows an orbit that takes it out as far as 65 thousand light years from the Galactic center and then back in as close as about 14 thousand light years, on a path inclined to the Galactic disk (by about 30°), taking about 4 hundred million years to make one revolution around our Galaxy.

M 80 (NGC 6093)

Constellation	Object type	RA, Dec	Approx. transit date at local midnight	Distance
Scorpius	Globular cluster	16h 17.0m, −22° 59′	June 21	30 thousand light years
Age	Apparent size	Magnitude	Sky Atlas 2000.0 chart	Herald–Bobroff chart
12–14 billion years	5.1′	7.3	22	C63/D13

It has a mass of nearly 4 hundred thousand Suns and a diameter of about 50 light years. It was the first globular cluster to have a nova discovered in it. M 80 is a bulge cluster (meaning it orbits inside the central bulge of our Galaxy – see M 9/NGC 6333) and has one of the shortest orbital periods of the globular clusters in our Galaxy. Indeed, it only takes about 70 million years to complete one revolution about the Galaxy, in an orbit that is highly inclined to the Galactic central plane.

M 81 (NGC 3031)

Constellation	Object type	RA, Dec	Approx. transit date at local midnight	Distance
Ursa Major	Spiral galaxy	09h 55.6m, +69° 04′	February 28	13 million light years
Age	Apparent size	Magnitude	Sky Atlas 2000.0 chart	Herald–Bobroff chart
	24.9′ × 11.5′	6.9	2	C13

M 81 is a LINER (low-ionization nuclear emission region) galaxy, which is a low-luminosity class of "active galactic nuclei" (AGN). The mechanism for LINERs remains uncertain but in some galaxies it may be due to a supermassive black hole in the nucleus that is accreting gas and stars, resulting in photoionization of surrounding gas. (Indeed, some suggest that LINER galaxies represent an evolutionary stage between quasars and ordinary galaxies.) Alternatively, some LINERs may instead be caused by intense star-formation activity in the nucleus (a "starburst"). M 81 is thought to be in the former class (a LINER whose emission is associated with a central black hole), with its supermassive black hole estimated to contain about 60 million solar masses. About one third of all galaxies are LINERs. M 81 is the namesake member of the M 81 group of at least 11 gravitationally bound galaxies that includes NGC 2403 (14° W), NGC 2976 (1° 23′ SW), IC 2574 (3° E), NGC 4236 (12° E), in addition to nearby M 82/NGC 3034 and NGC 3077 with which M 81 has had strong past interactions (see NGC 3077 and M 82/NGC 3034). M 81 has an optical diameter of about 95 thousand light years and a mass of about 50 billion Suns.

M 82 (NGC 3034)

Constellation	Object type	RA, Dec	Approx. transit date at local midnight	Distance
Ursa Major	Irregular galaxy	09h 55.9m, +69° 41'	February 28	13 million light years

Age	Apparent size	Magnitude	Sky Atlas 2000.0 chart	Herald–Bobroff chart
	10.5' × 5.1'	8.4	2	C14

This is the prototypical starburst galaxy (in which intense star formation is occurring in its central region). Indeed, three times as many stars are forming every year in the center of this galaxy (within a radius of about 0.5' of this galaxy's center) as form in the entire Milky Way in one year. Supernovae occur in this starburst region about once a decade (which is several times the rate for the entire Milky Way). These supernovae blow material out of the center of this galaxy in a superwind (moving at 5–8 hundred km/s) that forms two jets perpendicular to the plane of the galaxy. These jets (which pick up material on their way out, possibly by turbulent shear layer mixing and by evaporating nearby gas in the galaxy), are believed to be slamming into gas in a halo outside the galaxy's disk. This halo gas is thought to be left over from earlier gravitational interactions with nearby M 81. Material in the superwind jets is thought to be moving faster than the escape velocity of the galaxy and so will become intergalactic material. M 82 is part of the M 81 group of gravitationally bound galaxies (see M 81/NGC 3031) and is thought to have had strong interactions with M 81 over the last several hundred million years that have triggered the starburst in M 82. Associated with the bright star-forming regions, over one hundred "super star clusters" are known in M 82, thought to be newly minted globular clusters containing millions of Suns each. The optical diameter of M 82 is about 40 thousand light years.

M 83 (NGC 5236)

Constellation	Object type	RA, Dec	Approx. transit date at local midnight	Distance
Hydra	Barred spiral galaxy	13h 37.0m, −29° 52'	May 11	12 million light years

Age	Apparent size	Magnitude	Sky Atlas 2000.0 chart	Herald–Bobroff chart
	13.1' × 12.2'	7.5	21	C65/D29

This is a starburst galaxy (meaning it has intense star formation occurring – see M 82/NGC 3034) with the star formation concentrated in a half-ringlet occupying the region 3–7" from the galaxy center that contains hundreds of star clusters. About thirty of these star clusters have masses of more than 20 thousands Suns and are less than 10 million years old. M 83 is part of a gravitationally bound group of galaxies that includes nearby NGC 5264 (1° E) and NGC 5253 (1° 53' SSE), the latter having its closest approach to M 83 one or 2 billion years ago. M 83 is currently in second place in the contest for the galaxy with the most recorded supernovae, having 6, which is second only to the 7 recorded in NGC 6946. In common with M 31, professional telescopes find M 83 may have a double nucleus (each containing about 130 million solar masses and separated by 2.7" at a position angle of 243° i.e. aligned along a WSW–ENE direction). M 83 has an optical diameter of about 50 thousand light years and is nearly face-on (with an inclination angle of 24° i.e. it rotates about an axis that is inclined from our line-of-sight by 24°).

M 84 (NGC 4374)

Constellation	Object type	RA, Dec	Approx. transit date at local midnight	Distance
Virgo	Elliptical galaxy	12h 25.1m, +12° 53'	April 23	55 million light years
Age	**Apparent size**	**Magnitude**	**Sky Atlas 2000.0 chart**	**Herald–Bobroff chart**
	6.7' × 6.0'	9.1	14	C48/D33/E10

It is part of the Virgo galaxy cluster (see M 49/NGC 4472) and a physically close companion to M 86. M 84 lies at one end of the "Markarian Chain" of galaxies that lies along a "chain" NE of M 84 and includes eight galaxies: M 84 (NGC 4374), M 86 (NGC 4406), NGC 4435, NGC 4438, NGC 4461, NGC 4458, NGC 4473 and NGC 4477. These galaxies are moving like a rigid, tumbling chain thrown away from Earth at several hundred km/s with the chain tumbling so that the W side of the chain (M 84) is actually moving toward us while the E side (NGC 4477) is moving doubly fast away from us. The nucleus of M 84 is an AGN (or "active galactic nucleus"), in which energetic emission is caused by accretion onto a massive central object, which for M 84 is thought to be a black hole with a mass of several hundred million Suns. This AGN powers the two jets and associated lobes that are evident in the nuclear region (the inner few arcseconds) in professional telescopes at radio wavelengths, the radio waves being due to synchrotron emission of high-speed electrons gyrating wildly in a magnetic field. M 84 is one of fewer than 20 galaxies to have had three or more recorded supernovae.

M 85 (NGC 4382)

Constellation	Object type	RA, Dec	Approx. transit date at local midnight	Distance
Coma Berenices	Lenticular galaxy	12h 25.4m, +18° 11'	April 23	55 million light years
Age	**Apparent size**	**Magnitude**	**Sky Atlas 2000.0 chart**	**Herald–Bobroff chart**
	7.4' × 5.9'	9.1	14	C30/D33/E10

M 85 is at the northern edge of the Virgo galaxy cluster (see M 49/NGC 4472) and a close physical companion to NGC 4394 (7' ENE). It is viewed nearly face-on and is unusual for a lenticular galaxy because it contains young stars (i.e. less than a few billion years old) in its inner regions, which may have arisen from a past merger with another galaxy or because of interaction with nearby NGC 4394.

M 86 (NGC 4406)

Constellation	Object type	RA, Dec	Approx. transit date at local midnight	Distance
Virgo	Elliptical galaxy	12h 26.2m, +12° 57′	April 23	55 million light years
Age	Apparent size	Magnitude	Sky Atlas 2000.0 chart	Herald–Bobroff chart
	9.8′ × 6.3′	8.9	14	C48/D33/E10

It is part of the Virgo galaxy cluster (see M 49/NGC 4472), lying toward the back side of the main concentration of the Virgo cluster (i.e. lying perhaps a little less than 10 million light years farther away than M 87). Along with its close physical companion M 84, it is part of the "Markarian Chain" of galaxies (see M 84/NGC 4374). M 86 is moving rapidly toward us compared to the rest of the Virgo cluster, resulting in M 86 having a high speed (over a thousand km/s) relative to the material between galaxies in the cluster (the "intracluster medium"). This galaxy's high-speed movement through the intracluster medium is thought to be causing gas to be stripped from the galaxy (so-called "ram-pressure stripping").

M 87 (NGC 4486)

Constellation	Object type	RA, Dec	Approx. transit date at local midnight	Distance
Virgo	Elliptical galaxy	12h 30.8m, +12° 23′	April 24	55 million light years
Age	Apparent size	Magnitude	Sky Atlas 2000.0 chart	Herald–Bobroff chart
	8.7′ × 6.6′	8.6	14	C48/D33/E09

Lying near the center of the Virgo galaxy cluster (see M 49/NGC 4472), this is a heavy-weight in the world of galaxies. Its exact mass is uncertain but is probably several trillion solar masses (although much of this is dark matter). This is many hundreds of times more massive than the average galaxy. M 87 lies at the heart of the largest of the two major subclusters within the Virgo cluster (the other being associated with M 49/NCG 4472), with the M 87 clump of the Virgo cluster having a mass of several hundred trillion Suns. The center of M 87 is thought to contain a supermassive black hole with a mass of several billion Suns. This black hole is believed to be what drives the active galactic nucleus (in which energetic emission is caused by accretion onto the massive central black hole) that gives rise to the optically one-sided jet in M 87. This jet is the major part of the radio source known as Virgo A, the radio waves being due to synchrotron emission of high-speed electrons gyrating wildly in a magnetic field. This jet extends for thousands of light years (its brightest parts extending out about 25″), with material in the jet travelling at significant fractions of the speed of light. M 87 has one of the largest numbers of globular clusters of any galaxy, with more than 13 thousand (compare to the Milky Way's 150 or so). The optical diameter of M 87 is about 140 thousand light years (which is more than twice the diameter of the average galaxy), but it extends out several times this distance in professional telescopic studies.

M 88 (NGC 4501)

Constellation	Object type	RA, Dec	Approx. transit date at local midnight	Distance
Coma Berenices	Spiral galaxy	12h 32.0m, +14° 25′	April 24	55 million light years

Age	Apparent size	Magnitude	Sky Atlas 2000.0 chart	Herald–Bobroff chart
	6.8′ × 3.7′	9.6	14	C48/D33/E09

M 88 is part of the Virgo galaxy cluster (see M 49/NGC 4472) and is a Seyfert galaxy (where a supermassive object in this galaxy's center accumulates nearby gas resulting in strong emission from the nucleus). It has a mass of about 1/4 billion Suns and an optical diameter of about 110 thousand light years. The nucleus of this galaxy (within a radius of about 4″) has a different chemical makeup than the rest of the galaxy. The galaxy is a "flocculent" spiral, meaning that it lacks any obvious azimuthally symmetric spiral arm pattern (see NGC 3521).

M 89 (NGC 4552)

Constellation	Object type	RA, Dec	Approx. transit date at local midnight	Distance
Virgo	Elliptical galaxy	12h 35.7m, +12° 33′	April 24	55 million light years

Age	Apparent size	Magnitude	Sky Atlas 2000.0 chart	Herald–Bobroff chart
	5.3′ × 4.8′	9.8	14	C48/D33/E09

It is part of the Virgo galaxy cluster (see M 49/NGC 4472). A supermassive object in this galaxy's center (thought to be a black hole with a mass of many hundreds of millions of Suns) accumulates nearby gas resulting in emission from the nucleus, making it a weak Seyfert galaxy or else a LINER ("low-ionization nuclear emission region" – see M 81/NGC 3031) galaxy.

M 90 (NGC 4569)

Constellation	Object type	RA, Dec	Approx. transit date at local midnight	Distance
Virgo	Barred spiral galaxy	12h 36.8m, +13° 10′	April 26	55 million light years

Age	Apparent size	Magnitude	Sky Atlas 2000.0 chart	Herald–Bobroff chart
	9.9′ × 4.4′	9.5	14	C48/D33/E09

It is part of the Virgo cluster (see M 49/NGC 4472). It is a LINER galaxy ("low-ionization nuclear emission region" – see M 81/NGC 3031), but its nuclear emission is thought to be dominated by intense star formation (i.e. a starburst) and not by accretion onto a central black hole. About two new stars form every year in this galaxy (about half the number forming yearly in the Milky Way). The galaxy rotates about an axis that is inclined to our line-of-sight by about 64° with the western edge of the galaxy closest to us. Although the galaxy IC 3583 (6′ NNW) is nearby in the sky and is also a Virgo cluster member, the two are not thought to be interacting strongly. Because of M 90's high speed relative to the Virgo cluster (over a thousand km/s), it is thought to have lost some of its gas to "ram-pressure stripping" (see M 91/NGC 4548).

M 91 (NGC 4548)

Constellation	Object type	RA, Dec	Approx. transit date at local midnight	Distance
Coma Berenices	Barred spiral galaxy	12h 35.4 m, +14° 30′	April 24	55 million light years
Age	**Apparent size**	**Magnitude**	**Sky Atlas 2000.0 chart**	**Herald–Bobroff chart**
	5.2′ × 4.2′	10.1	14	C48/D33/E09

This galaxy is part of the M 87 clump of the Virgo galaxy cluster (see M 49/NGC 4472). It is a LINER galaxy (see M 81/NGC 3031). Like many spirals in galaxy clusters, this galaxy has much less atomic hydrogen gas than average for spirals in general. This is thought to be caused by stripping of this gas (so-called "ram-pressure stripping") from the galaxy due to the galaxy's motion relative to the material between galaxies in the cluster (the "intracluster medium"). Most of this galaxy rotates at about 250 km/s about an axis inclined to our line-of-sight by 40°. It is thought that a mistake was made by Messier in listing the position of this object, leading to confusion as to which object M 91 referred to, until the probable origin of Messier's error was uncovered in 1969 by W. C. Williams.

M 92 (NGC 6341)

Constellation	Object type	RA, Dec	Approx. transit date at local midnight	Distance
Hercules	Globular cluster	17h 17.1m, +43° 08′	July 6	27 thousand light years
Age	**Apparent size**	**Magnitude**	**Sky Atlas 2000.0 chart**	**Herald–Bobroff chart**
12–14 billion years	12.2′	6.4	8	C26

This is one of the oldest globular clusters known and is very metal-poor (i.e. having a low abundance of elements heavier than helium). It has a mass of about 4 hundred thousand Suns. Its size of 12.2′ corresponds to a diameter of about one hundred light years. It orbits the Galaxy on a path that is somewhat inclined to the disk (by a little more than 20°) that takes it a maximum of about 13 thousand light years away from the Galactic central plane, travelling between about 5 and 35 thousand light years from the Galactic center. It completes one revolution around the Galaxy about once every 2 hundred million years.

M 93 (NGC 2447)

Constellation	Object type	RA, Dec	Approx. transit date at local midnight	Distance
Puppis	Open cluster	07h 44.5m, −23° 51′	January 26	3.5 thousand light years
Age	**Apparent size**	**Magnitude**	**Sky Atlas 2000.0 chart**	**Herald–Bobroff chart**
4 hundred million years	22′	6.2	19	C69/D21

It lies just above the Galactic central plane (by a few light years) a few thousand light years farther from the Galactic center than we are. It has a diameter of about 20 light years.

M 94 (NGC 4736)

Constellation	Object type	RA, Dec	Approx. transit date at local midnight	Distance
Canes Venatici	Spiral galaxy	12h 50.9 m, +41° 07′	April 29	15 million light years
Age	**Apparent size**	**Magnitude**	**Sky Atlas 2000.0 chart**	**Herald–Bobroff chart**
	12.3′ × 10.8′	8.2	7	C29/D30

It has a mass of about 60 billion Suns and an optical diameter of about 50 thousand light years. Professional telescopes find M 94 has an inner ring (45″ in radius), as well as an outer ring (5.5′ in radius). The inner ring is undergoing intense star formation (i.e. it is a "starburst" ring), creating about 1.5 new stars/year. The rings are thought to be the result of orbital resonant interactions where density waves are in resonance with the local speed of epicycle oscillations in the orbits of matter in the disk, including the so-called Lindblad resonances. Most of the galaxy rotates at approximately 150 km/s about an axis inclined to our line-of-sight by about 40°. M 94 is a LINER galaxy (see M 81/NGC 3031) and is part of the M 106 group of 17 gravitationally bound galaxies (see M 106/NGC 4258).

M 95 (NGC 3351)

Constellation	Object type	RA, Dec	Approx. transit date at local midnight	Distance
Leo	Barred spiral galaxy	10h 44.0 m, +11° 42′	March 12	35 million light years
Age	**Apparent size**	**Magnitude**	**Sky Atlas 2000.0 chart**	**Herald–Bobroff chart**
	7.3′ × 4.4′	9.7	13	C49/D36

It is part of the M 96 (NGC 3368) galaxy group – see M 96/NGC 3368. Professional telescopes find that M 95 has an inner ring (7″ in radius) and an outer ring (1′ in diameter), both being ionized hydrogen (HII) star-forming regions and due to orbital resonant interactions with this galaxy's bar. M 95 is a nearly average galaxy in size and mass, having a mass of about 50 billion Suns and an optical diameter of about 70 thousand light years.

M 96 (NGC 3368)

Constellation	Object type	RA, Dec	Approx. transit date at local midnight	Distance
Leo	Barred spiral galaxy	10h 46.8 m, +11° 49′	March 13	35 million light years
Age	**Apparent size**	**Magnitude**	**Sky Atlas 2000.0 chart**	**Herald–Bobroff chart**
	7.8′ × 5.2′	9.3	13	C49/D36

It is the namesake member of the M 96 (NGC 3368) group of 12 galaxies that includes NGC 3299, NGC 3351 (M 95), NGC 3377 (see NGC 3377), M 105 (NGC 3379), NGC 3384 (see NGC 3384), NGC 3412 (see NGC 3412), and NGC 3489 (see NGC 3489). M 96 has a mass of about 80 billion Suns, an optical diameter of about 80 thousand light years and is a LINER (see M 81/NGC 3031).

M 97 (NGC 3587)

Constellation	Object type	RA, Dec	Approx. transit date at local midnight	Distance
Ursa Major	Planetary nebula	11h 14.8 m, +55° 01'	March 20	2 thousand light years
Age	**Apparent size**	**Magnitude**	**Sky Atlas 2000.0 chart**	**Herald–Bobroff chart**
	3.2'	11	2	C12/D35

Nicknamed the "Owl Nebula" because it resembles two owl's eyes in a round disk when viewed at high power in large amateur telescopes. The nebula is expanding at a few tens of km/s and has a mass about 13% that of our Sun (not including the central star, whose mass is about 60% that of the Sun and is a challenging object in amateur telescopes). The three-dimensional structure of this nebula is complex in professional telescopic studies, although the "owl's eyes" are thought to be part of a bipolar ("hourglass") barrel-shaped structure (inclined from our line-of-sight by about 45°) that is inside three separate elliptical shells. The inner shell of the three gives rise to the round outer shape visible in amateur telescopes.

M 98 (NGC 4192)

Constellation	Object type	RA, Dec	Approx. transit date at local midnight	Distance
Coma Berenices	Barred spiral galaxy	12h 13.8 m, +14° 54'	April 20	55 million light years
Age	**Apparent size**	**Magnitude**	**Sky Atlas 2000.0 chart**	**Herald–Bobroff chart**
	9.4' × 2.3'	10.1	14	C48/D33/E10

M 98 is part of the Virgo galaxy cluster (see M 49/NGC 4472), near its western edge and thought to be lying toward the front of this cluster. It has a mass of about 2 hundred billion Suns and an optical diameter of about 150 thousand light years. It is nearly edge-on (with its axis of rotation inclined by about 80° from our line-of-sight). It has emission from its nucleus (perhaps from a LINER that is powered by both a starburst and accretion onto a massive central object – see M 81/NGC 3031).

M 99 (NGC 4254)

Constellation	Object type	RA, Dec	Approx. transit date at local midnight	Distance
Coma Berenices	Spiral galaxy	12h 18.8 m, +14° 25'	April 21	55 million light years
Age	**Apparent size**	**Magnitude**	**Sky Atlas 2000.0 chart**	**Herald–Bobroff chart**
	5.3' × 4.6'	9.9	14	C48/D33/E10

This is one of several galaxies whose nickname is the "Pinwheel Galaxy" (due to its multiple spiral arms evident in professional telescopic images). It is part of the Virgo galaxy cluster (see M 49/NGC 4472). It has a mass of a little over a hundred billion Suns and an optical diameter of about 90 thousand light years. Vigorous star formation is occurring throughout this galaxy. In professional telescopes, one of this galaxy's spiral arms is much more pronounced, possibly caused by a large cloud of gas (with a mass of hundreds of millions of Suns) that is falling into the galaxy and which may be the leftover scraps of a galaxy that M 99 previously disrupted.

M 100 (NGC 4321)

Constellation	Object type	RA, Dec	Approx. transit date at local midnight	Distance
Coma Berenices	Barred spiral galaxy	12h 22.9 m, +15° 49′	April 22	55 million light years
Age	**Apparent size**	**Magnitude**	**Sky Atlas 2000.0 chart**	**Herald–Bobroff chart**
	7.5′ × 6.1′	9.4	14	C48/D33/E10

It is part of the Virgo galaxy cluster (see M 49/NGC 4472), being the brightest and largest spiral galaxy in the Virgo cluster. It is a "grand-design spiral", meaning that it has two symmetrically placed spiral arms that extend over most of its visible disk in professional telescopic images. Resonance associated with the bar and the "Lindblad resonances" (where the speed of density waves resonantly amplifies oscillations in the orbits of matter in the disk) are thought to be triggering a ring of star formation (a "starburst") in the central region of M 100. This ring has a radius of 7.5–20″ and is perhaps 10 million years old. M 100 is one of fewer than 10 galaxies that has had four or more recorded supernovae. M 100 has a mass of about 2 hundred billion Suns and an optical diameter of about 120 thousand light years.

M 101 (NGC 5457)

Constellation	Object type	RA, Dec	Approx. transit date at local midnight	Distance
Ursa Major	Barred spiral galaxy	14h 03.2 m, +54° 21′	May 17	25 million light years
Age	**Apparent size**	**Magnitude**	**Sky Atlas 2000.0 chart**	**Herald–Bobroff chart**
	28.5′ × 28.3′	7.9	2	C11

This galaxy is nearly face-on. It has a diameter of almost 2 hundred thousand light years. Many ionized hydrogen (HII) star-forming regions are present in M 101, some of which can be seen as bright knots in amateur telescopes. Several of these knots have their own NGC numbers e.g. NGC 5461, 5462, 5471, the latter of which is a hundred times larger and brighter than any HII region in our Galaxy. Some of these supersized HII regions have masses of tens of millions of Suns (e.g. NGC 5461 and 5471), and although they appear as a single knot in amateur telescopes, they are made up of many individual giant molecular clouds (GMCs) that have masses of several hundred thousands of Suns (similar to the masses of GMCs in our Galaxy). M 101 is the namesake member of the M 101 group of gravitationally bound galaxies that includes NGC 5474 (44′ SSE), NGC 5585 (3° 21′ NE), NGC 5204 (6° NW) and NGC 5477 (22′ ENE) and Holmberg IV (UGC 8837, 1° 19′ WSW). Tidal interactions with several group members in the past few hundred million years (perhaps including NGC 5477, NGC 5474 and Holmberg IV) are thought to have distorted M 101. These interactions may have induced the formation of some of the HII regions e.g. NGC 5471.

M 102 (NGC 5866)

Constellation	Object type	RA, Dec	Approx. transit date at local midnight	Distance
Draco	Lenticular galaxy	15h 06.5m, +55° 46′	June 2	50 million light years
Age	**Apparent size**	**Magnitude**	**Sky Atlas 2000.0 chart**	**Herald–Bobroff chart**
	6.5′ × 3.1′	9.9	2	C09/C10

This is a lenticular galaxy (given the label S0 in classification schemes), meaning it has a disk and central bulge like a spiral galaxy, but lacks the spiral arms. Lenticulars contain very little gas, dust, or young stars, consisting almost entirely of old stars. This galaxy is sometimes called the "Spindle Galaxy", although NGC 3115 also has this nickname. It is nearly edge-on (inclination angle of 71°, meaning it rotates about an axis that is inclined at 71° from our line-of-sight) with about 60% of its light coming from its bulge and 40% from its disk. It has an optical diameter of 90 thousand light years (the diameter of our Galaxy is about 1 hundred thousand light years). It is the namesake member of the NGC 5866 group of galaxies that includes nearby NGC 5907 and NGC 5879, which are within several million light years of NGC 5866. Messier's original M 102 is believed to be a duplicate entry of M 101 (NGC 5457) rather than being NGC 5866, but in order to have 110 different objects in the Messier list, many amateur astronomers informally attach the label M 102 to NGC 5866.

The Messier Objects

M 103 (NGC 581)

Constellation	Object type	RA, Dec	Approx. transit date at local midnight	Distance
Cassiopeia	Open cluster	01h 33.4m, +60° 39′	October 24 (Standard Time)	9 thousand light years
Age	**Apparent size**	**Magnitude**	**Sky Atlas 2000.0 chart**	**Herald–Bobroff chart**
20 million years	6.0′	7.4	1	C20/D27

Of 228 stars with magnitude 14.5 and brighter that have been examined in professional telescopic studies in the cluster region, 77 are known to be actual cluster members, the rest being field stars (see M 39/NGC 7092 for further discussion). The brightest star in the cluster region (a double star, Struve 131, mag. 7.3 and 10.5 with separation 13.8″ along a SE–NW direction) is not a cluster member, but is a field star (in the foreground). M 103's size of 6′ corresponds to a diameter of about 15 light years.

M 104 (NGC 4594)

Constellation	Object type	RA, Dec	Approx. transit date at local midnight	Distance
Virgo	Spiral galaxy	12h 40.0 m, −11° 37′	April 25	30 million light years
Age	Apparent size	Magnitude	Sky Atlas 2000.0 chart	Herald– Bobroff chart
	8.6′ × 4.2′	8.0	14	C48/D31

Nicknamed the "Sombrero Galaxy". The well-known dust lane, which actually has a ring-shape (like Saturn's rings), contains about 10 million solar masses worth of dust grains that are submicron in size. The dust lane may owe its existence to the gravitational interaction of a now-defunct bar with the interstellar medium. The disk in this galaxy is about 1/4 as massive as this galaxy's large spheroidal bulge (whereas in our Galaxy the disk has a mass about 7 times that of the bulge). Together the disk and bulge have a mass of several hundred billion Suns. This galaxy harbors a central black hole with a mass of about a billion Suns, onto which a few per cent of a solar mass is accreted every year. It is either a LINER (see M 81/NGC 3031) or Seyfert (see NGC 3227) galaxy, is nearly edge-on (inclination angle of 84°) and contains well over a thousand globular clusters.

M 105 (NGC 3379)

Constellation	Object type	RA, Dec	Approx. transit date at local midnight	Distance
Leo	Elliptical galaxy	10h 47.8 m, +12° 35′	March 13	35 million light years
Age	Apparent size	Magnitude	Sky Atlas 2000.0 chart	Herald– Bobroff chart
	5.3′ × 4.8′	9.3	13	C31/D36

It is part of the M 96 (NGC 3368) group of galaxies (see M 96/NGC 3368). It has a mass of about one hundred billion Suns. It contains a supermassive central dark mass (thought to be a black hole) with a mass of one or 2 hundred million Suns. NGC 3389 (spiral galaxy, 10′ ESE, mag. 11.9, with a mass similar to M 105) is nearby, although it is actually a background object and is instead about twice as far away and part of the NGC 3338 group of galaxies that includes NGC 3338, NGC 3389 and NGC 3346.

M 106 (NGC 4258)

Constellation	Object type	RA, Dec	Approx. transit date at local midnight	Distance
Canes Venatici	Barred spiral galaxy	12h 19.0 m, +47° 18′	April 20	25 million light years
Age	Apparent size	Magnitude	Sky Atlas 2000.0 chart	Herald–Bobroff chart
	17.4′ × 6.6′	8.4	7	C12/D35

This is a LINER galaxy (see M 81/NGC 3031) and is thought to harbor a central black hole with a mass of about 40 million Suns that is accreting a mass of perhaps 0.01 Suns per year. It is the nearest galaxy to us that has extragalactic astrophysical jets. These bipolar jets, driven by the central black hole, are thought to emanate from the nucleus at about a 30° angle from the plane of the disk and cause shocks in the gas of the disk which appear as "anomalous arms" in professional telescopic studies. The nucleus of this galaxy also contains water masers (see NGC 3079). This galaxy has an optical diameter of about 130 thousand light years. It is at the north end of the NGC 4258 (M 106) namesake group of 17 gravitationally bound galaxies that includes NGC 4144 (1° 45′ WSW), NGC 4242 (1° 42′ S), NGC 4460 (3° SE), NGC 4490 (6° SSE – see NGC 4490), NGC 4485 (6° SSE – see NGC 4485), NGC 4618 (7° SE – see NGC 4618), NGC 4625 (7° SE), NGC 4449 (3.5° SSE – see NGC 4449) and possibly NGC 4248 (14′ WNW).

M 107 (NGC 6171)

Constellation	Object type	RA, Dec	Approx. transit date at local midnight	Distance
Ophiuchus	Globular cluster	16h 32.5m, −13° 03′	June 23	20 thousand light years
Age	Apparent size	Magnitude	Sky Atlas 2000.0 chart	Herald–Bobroff chart
12–14 billion years	3.3′	8.1	15	C45

It has a mass of about 2 hundred thousand Suns and a diameter of about 20 light years. It is a bulge cluster (i.e. it orbits in the central ball-like bulge of our Galaxy) with a period of about one hundred million years. Its orbital path is inclined by about 45° to the Galactic disk and is a very flattened ellipse. It was not noted by Messier, but instead is a recent (1947) addition to the Messier catalogue suggested by H.S. Hogg.

M 108 (NGC 3556)

Constellation	Object type	RA, Dec	Approx. transit date at local midnight	Distance
Ursa Major	Barred spiral galaxy	11h 11.5 m, +55° 40'	March 19	40 million light years
Age	Apparent size	Magnitude	Sky Atlas 2000.0 chart	Herald–Bobroff chart
	8.6' × 2.4'	10.0	2	C13/D35

Professional telescopic studies find that this galaxy has two giant loops of atomic hydrogen gas, one at the east and one at the west end. These loops have diameters of 10–20 thousand light years, have masses around 50 million Suns and are expanding outward but parallel to the disk of the galaxy at 40–50 km/s. They are thought to have originated about 50 million years ago. The loops may be the result of the rapid expansion of jets shot outward from an active galactic nucleus (powered by accretion onto a supermassive central black hole), with the jets "flaring" into shells as they reach the less dense outer regions of the galaxy. The nuclear activity has since largely subsided (since such activity is thought to last only some tens of millions of years in spirals). The galaxy is nearly edge-on (having an inclination angle of about 75°, meaning that it rotates about an axis inclined at 75° from our line-of-sight) and has an optical diameter of about 1 hundred thousand light years.

M 109 (NGC 3992)

Constellation	Object type	RA, Dec	Approx. transit date at local midnight	Distance
Ursa Major	Barred spiral galaxy	11h 57.6 m, +53° 22'	March 31	60 million light years
Age	Apparent size	Magnitude	Sky Atlas 2000.0 chart	Herald–Bobroff chart
	7.5' × 4.4'	9.8	2	C12/D35

It has a mass of about 250 billion Suns and a diameter of about 130 thousand light years. It is part of the Ursa Major galaxy cluster, one of only three major galaxy clusters within 150 million light years of us (the others being the Virgo cluster and the Fornax cluster). Galaxy clusters are the largest stable structures in the Universe. Only about 1% of galaxies belong to galaxy clusters. The Ursa Major cluster contains about 80 known galaxies, and has about 5% of the mass (but emits 30% of the light) of the Virgo cluster (see M 49/NGC 4472). The Ursa Major cluster is an unusual cluster in that it consists almost entirely of late-type galaxies (e.g. Sc and SBc and later galaxies in Hubble's galaxy classification scheme; Sc and SBc spirals are "late-type" spirals that have prominent, loosely wound, spiral arms and only a very small central bulge relative to an extended disk). In contrast, most galaxy clusters consist of "early-type" galaxies (i.e. elliptical and lenticular, and early-type spirals, the latter having relatively tightly wound spiral arms and a large central bulge). Indeed, 3/4 of the Virgo cluster galaxies are early-type. M 109 is the namesake member of the M 109/NGC 3992 galaxy group of gravitationally bound galaxies that is a subgroup of the Ursa Major cluster and consists of 27 galaxies.

M 110 (NGC 205)

Constellation	Object type	RA, Dec	Approx. transit date at local midnight	Distance
Andromeda	Elliptical galaxy	00h 40.4m, +41° 41′	October 26	2.2 million light years

Age	Apparent size	Magnitude	Sky Atlas 2000.0 chart	Herald– Bobroff chart
	20′ × 12′	8.1	4	C38

This is a dwarf galaxy (about 10 thousand light years in optical diameter, and having a mass of about 10 billion Suns). It is in our Local Group and close to M 31. The mass of interstellar gas in this galaxy, 1 million Suns worth, is about 10 times less than expected and is rotating, while the stellar component of this galaxy is not rotating. These unusual features may be due to a past interaction with M 31, which might also explain the recent burst of star formation in NGC 205 that began 0.5 billion years ago and ended a few million years ago.

3

NGC (New General Catalogue) Objects

NGC 40

Constellation	Object type	RA, Dec	Approx. transit date at local midnight	Distance
Cepheus	Planetary nebula	00h 13.0m, +72° 31′	October 19	3 thousand light years
Age	**Apparent size**	**Magnitude**	**Sky Atlas 2000.0 chart**	**Herald–Bobroff chart**
	0.6′	11	1	C03

It has an actual diameter of about 1/2 light year. The bright central star (mag. 11.5) has a mass of about 0.7 Suns and is a Wolf–Rayet star. Wolf–Rayet stars are relatively rare with only about 230 known in our Galaxy; about 50 of these reside as central stars in planetary nebulae so that about 10% of the central stars in planetary nebulae are Wolf–Rayet stars. Wolf–Rayet stars are very hot (around 1 hundred thousand K), ejecting mass at a high-speed (several thousand km/s) and in the last evolutionary stages of their life. Wolf–Rayet stars in planetary nebulae are of the "WC" type, meaning they are rich in carbon (as opposed to the other main kind of Wolf–Rayets, "WN", which are rich in nitrogen).

NGC 129

Constellation	Object type	RA, Dec	Approx. transit date at local midnight	Distance
Cassiopeia	Open cluster	00h 30.0m, +60° 13′	October 23	7 thousand light years
Age	**Apparent size**	**Magnitude**	**Sky Atlas 2000.0 chart**	**Herald–Bobroff chart**
80 million years	21′	6.5	1	C03/D06

It has an actual diameter of about 40 light years, making it a relatively large open cluster in actual size. Nearly 4 hundred stars have been counted as members of this cluster in professional telescopic studies. The Cepheid variable DL Cas can be seen as the 8.9 mag. star that is the northern of the three bright mag. 9 members that form a triangle (with 3′ long sides) near the center cluster (see NGC 7790 for a brief explanation of Cepheid variable stars). DL Cas has a mass of 5.6 Suns, a radius 66 times that of our Sun, and is a spectroscopic binary.

NGC 136

Constellation	Object type	RA, Dec	Approx. transit date at local midnight	Distance
Cassiopeia	Open cluster	00h 31.5m, +61° 31′	October 24	13 thousand light years
Age	**Apparent size**	**Magnitude**	**Sky Atlas 2000.0 chart**	**Herald–Bobroff chart**
2 hundred million years	1.2′	11.5	1	C03/D06

It has a diameter of about 5 light years and was discovered in 1788 by William Herschel.

NGC 157

Constellation	Object type	RA, Dec	Approx. transit date at local midnight	Distance
Cetus	Barred spiral galaxy	00h 34.8m, –08° 24′	October 25 (Daylight Savings Time)	70 million light years
Age	**Apparent size**	**Magnitude**	**Sky Atlas 2000.0 chart**	**Herald–Bobroff chart**
80 million years	4.1′ × 2.7′	10.4	10	C39

It has an outer, slow-rotating, gas disk (with rotational velocity near 125 km/s). This outer disk surrounds a more rapidly spinning inner disk which is 2′ in diameter and "warped" so that the inner region lies in a plane that is inclined by about 10° from the outer disk. The outer disk rotates about an axis that is inclined to our line-of-sight by approximately 50°.

NGC 185

Constellation	Object type	RA, Dec	Approx. transit date at local midnight	Distance
Cassiopeia	Elliptical galaxy	00h 39.0m, +48° 20′	October 26 (Daylight Savings Time)	2.2 million light years
Age	**Apparent size**	**Magnitude**	**Sky Atlas 2000.0 chart**	**Herald–Bobroff chart**
	12.5′ × 10.4′	9.2	4	C03

This is a dwarf galaxy (8 thousand light years in optical diameter) in our Local Group and close to M 31, lying 1/4 million light years away from it. Despite its small actual size, its closeness to us gives it a relatively large size. Like all dwarf elliptical galaxies, NGC 185 was thought to consist only of very old (Population II) stars. However, recent studies show star formation in this galaxy occurred as recently as one hundred million years ago, but only in a central region of apparent diameter 2′. Outside this region the stars are > 1 billion years old.

NGC 205 (M 110)

Constellation	Object type	RA, Dec	Approx. transit date at local midnight	Distance
Andromeda	Elliptical galaxy	00h 40.4m, +41° 41'	October 26	2.2 million light years
Age	**Apparent size**	**Magnitude**	**Sky Atlas 2000.0 chart**	**Herald–Bobroff chart**
	20' × 12'	8.1	4	C38

This is a dwarf galaxy (about 10 thousand light years in optical diameter, and having a mass of about 10 billion Suns). It is in our Local Group and close to M 31. The mass of interstellar gas in this galaxy, 1 million Suns worth, is about 10 times less than expected and is rotating, while the stellar component of this galaxy is not rotating. These unusual features may be due to a past interaction with M 31, which might also explain the recent burst of star formation in NGC 205 that began 0.5 billion years ago and ended a few million years ago.

NGC 221 (M 32)

Constellation	Object type	RA, Dec	Approx. transit date at local midnight	Distance
Andromeda	Elliptical galaxy	00h 42.7m, +40° 52'	October 26 (Daylight Savings Time)	2.4 million light years
Age	**Apparent size**	**Magnitude**	**Sky Atlas 2000.0 chart**	**Herald–Bobroff chart**
	8.5' × 6.5'	8.1	4	C38

This is a dwarf satellite galaxy of M 31 and is M 31's closest companion. M 32 is a very unusual galaxy, referred to as a "compact elliptical" galaxy, having an inordinately bright and compact central core. Its origin remains uncertain, but one hypothesis suggests it was once a spiral galaxy that was stripped down to its bulge a few billion years ago by intense tidal interaction with M 31. A black hole with a mass of several million Suns is thought to be present in the center of M 32. M 32 has an optical diameter of about 6 thousand light years.

NGC 224 (M 31)

Constellation	Object type	RA, Dec	Approx. transit date at local midnight	Distance
Andromeda	Spiral galaxy	00h 42.7m, +41° 16′	October 26 (Daylight Savings Time)	2.5 million light years
Age	**Apparent size**	**Magnitude**	**Sky Atlas 2000.0 chart**	**Herald–Bobroff chart**
	3.1° × 1.0°	3.4	4	C38

Nicknamed the "Andromeda Galaxy". This is the nearest large galaxy and is the most luminous member of our Local Group of galaxies (which contains about 40 galaxies in a radius of a few million light years). M 31 is thought to be roughly as massive as our Galaxy, but may be less massive (which would make our Galaxy the most massive member of the Local Group), although this remains uncertain. An apparent disk diameter of 3.1° corresponds to a diameter of 135 thousand light years, although the diameter visible in telescopes is larger than this (and depends on the aperture and observing conditions). Professional telescopes show it has perhaps as many as 15 dwarf spheroidal satellite galaxies, two of which [M 32 and NGC 205 (M 110)] are prominent in amateur telescopes. This is only a little more than the 11 such dwarf companions of our Galaxy. M 31 is very rich in globular clusters for a spiral galaxy, with nearly 5 hundred revealed in professional telescopes. The brightest of these globular clusters can be observed in amateur telescopes. Some of them are thought to have come from dwarf galaxies that M 31 accreted in past merger events. Professional telescopes find M 31 has a double nucleus containing a black hole with a mass of about 30 million Suns. The two nuclei are separated by 0.5″ and are thought to be part of an eccentric nuclear disk. The galaxy is a LINER ("low-ionization nuclear emission region" – see M 81/NGC 3031 for explanation).

NGC 225

Constellation	Object type	RA, Dec	Approx. transit date at local midnight	Distance
Cassiopeia	Open cluster	00h 43.6m, +61° 46′	October 27 (Daylight Savings Time)	2 thousand light years
Age	**Apparent size**	**Magnitude**	**Sky Atlas 2000.0 chart**	**Herald–Bobroff chart**
120 million years	12′	7.0	1	C03/D06

It contains about 30 stars in professional telescopic studies, but has a mass of about 70 Suns. It was discovered in 1784 by Caroline Herschel (William Herschel's sister).

NGC 246

Constellation	Object type	RA, Dec	Approx. transit date at local midnight	Distance
Cetus	Planetary nebula	00h 47.1m, −11° 52′	October 28 (Daylight Savings Time)	2 thousand light years
Age	**Apparent size**	**Magnitude**	**Sky Atlas 2000.0 chart**	**Herald–Bobroff chart**
·	3.8′	9	10	C56

Professional telescopes show that the central star (see Figure 246) is actually a binary system with a separation of 3.8″ and position angle of 130° (central star mag. 11.9, companion star mag. 14.3), with an orbital period of roughly 70 thousand years. The central star mass is rather high (0.8 Suns), suggesting the progenitor star had a mass about six times that of our Sun.

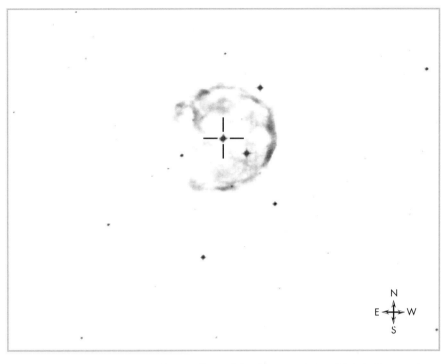

Figure 246 The cross-hairs indicate the central star of the planetary nebula NGC 246. The area shown is 15′ × 15′. From the Digitized Sky Survey (Space Telescope Science Institute) based on photographic data of the National Geographic Society – Palomar Observatory Sky Survey (POSS I).

NGC 247

Constellation	Object type	RA, Dec	Approx. transit date at local midnight	Distance
Cetus	Barred spiral galaxy	00h 47.1m, −20° 46′	October 28 (Daylight Savings Time)	8 million light years

Age	Apparent size	Magnitude	Sky Atlas 2000.0 chart	Herald–Bobroff chart
	21.0′ × 5.6′	9.1	18	C74

This is considered a dwarf spiral galaxy. Such galaxies have a much larger fraction of their mass in the halo compared to normal spirals, as well as having disk rotation velocities that increase outward throughout the disk (in contrast to normal spirals where most of the disk rotates at the same velocity). Star formation has occurred in the central regions of this galaxy within the last hundred million years. It is part of the Sculptor group of galaxies, a grouping that is about 3 million light years in diameter and contains a total of eight galaxies among which are NGC 55, 253, 300, and 7793. It has an optical diameter of about 50 thousand light years and a total dynamical mass of about 20 billion Suns.

NGC 253

Constellation	Object type	RA, Dec	Approx. transit date at local midnight	Distance
Sculptor	Barred spiral galaxy	00h 47.6m, –25° 17′	October 28 (Daylight Savings Time)	10 million light years
Age	**Apparent size**	**Magnitude**	**Sky Atlas 2000.0 chart**	**Herald– Bobroff chart**
	26.4′ × 6.0′	7.2	18	C57

This galaxy is part of the Sculptor group of galaxies (see NGC 247). Its rotation axis is inclined at 78° from our line-of-sight. It has active star formation occurring in its disk as well as a starburst nucleus (meaning violent, high-mass star formation is occurring there). Superwinds generated by the starburst nucleus are thought to be colliding with gas in the halo, thereby triggering star formation in the halo of this galaxy. A "super star cluster" is visible as a mag. 14 stellar point at the center of this galaxy, containing perhaps 1.5 million solar masses, and which may owe its origin to the accretion of a satellite body of 1 million solar masses that occurred 10 million years ago, and which may also have triggered the nuclear starburst in this galaxy.

NGC 278

Constellation	Object type	RA, Dec	Approx. transit date at local midnight	Distance
Cassiopeia	Barred spiral galaxy	00h 52.1m, +47° 33′	October 28 (Daylight Savings Time)	40 million light years
Age	**Apparent size**	**Magnitude**	**Sky Atlas 2000.0 chart**	**Herald– Bobroff chart**
	2.4′ × 2.4′	10.8	4	C03

The nucleus of this galaxy is dominated by stars younger than 50 million years, and star formation is still in process there. It has an optical diameter of about 30 thousand light years. Its rotation axis is inclined about 20–25° from our line-of-sight, meaning it is close to face-on.

NGC 281

Constellation	Object type	RA, Dec	Approx. transit date at local midnight	Distance
Cassiopeia	Emission nebula + open cluster	00h 53.0m, +56° 37′	October 29 (Daylight Savings Time)	10 thousand light years
Age	**Apparent size**	**Magnitude**	**Sky Atlas 2000.0 chart**	**Herald– Bobroff chart**
	4.0′	7.4	1	C20/D06

Nicknamed the "Pac-Man" nebula, from its resemblance to the character in the Pac-Man video arcade game that gained fame in the early 1980s (Pac-Man looks like a pizza with a slice missing). Like many emission nebulae (e.g. M 8 and M 16), this is a large ionized hydrogen (i.e. HII) star-forming region that is part of a much larger giant molecular cloud (GMC). In this case the GMC has a mass of almost 20 thousand Suns. NGC 281 is made visible by ionizing radiation from the young, hot stars embedded in it (which make up the open cluster IC 1590). Most of the ionizing is done by the bright star in the center of the nebula (HD 5005, mag. 7.4), which is a quadruple star of which three stars can be seen. The embedded cluster IC 1590 (which is unspectacular in amateur telescopes) is young (a few million years old) and contains 63 members brighter than mag. 17.

NGC 288

Constellation	Object type	RA, Dec	Approx. transit date at local midnight	Distance
Sculptor	Globular cluster	00h 52.8m, −26° 35′	October 29 (Daylight Savings Time)	30 thousand light years
Age	**Apparent size**	**Magnitude**	**Sky Atlas 2000.0 chart**	**Herald–Bobroff chart**
12–14 billion years	12.4′	8.1	18	C74

It lies within 1° of the South Galactic Pole i.e. it is located along a line nearly opposite to the direction of the axis of rotation of our Galaxy. It contains over 1 hundred thousand solar masses. It orbits the Galaxy on a path inclined by about 47° from the Galactic disk, taking a little over 2 hundred million years to complete an orbit that takes it out as far as 35 thousand light years from the Galactic center and as close in as 5 thousand light years from the Galactic center.

NGC 381

Constellation	Object type	RA, Dec	Approx. transit date at local midnight	Distance
Cassiopeia	Open cluster	01h 08.4m, +61° 35′	October 18 (Standard Time)	3 thousand light years
Age	**Apparent size**	**Magnitude**	**Sky Atlas 2000.0 chart**	**Herald–Bobroff chart**
320 million years	6′	9.3	1	C20/D27

It has a diameter of about 7 light years and was discovered in 1783 by Caroline Herschel (William Herschel's sister). Professional telescopic studies indicate it contains about 60 members. The eclipsing binary star OX Cas was once thought to belong to this cluster, but is now considered a field star.

NGC 404

Constellation	Object type	RA, Dec	Approx. transit date at local midnight	Distance
Andromeda	Lenticular galaxy	01h 09.4m, +35° 43′	October 18 (Standard Time)	8 million light years
Age	**Apparent size**	**Magnitude**	**Sky Atlas 2000.0 chart**	**Herald–Bobroff chart**
	4.3′ × 3.9′	10.3	4	C38/D42

This galaxy is unusual in its apparent lack of dark matter (which is normally present as an extended spherical halo beyond the visible regions of a galaxy). Practically all of NGC 404's several billion solar kinematical mass is contained in a 30 thousand light year diameter region. Active star formation is occurring only in the central nucleus. It has an inclination angle of about 10° to our line-of-sight.

NGC 436

Constellation	Object type	RA, Dec	Approx. transit date at local midnight	Distance
Cassiopeia	Open cluster	01h 16.0m, +58° 49′	October 19 (Standard Time)	8 thousand light years
Age	**Apparent size**	**Magnitude**	**Sky Atlas 2000.0 chart**	**Herald–Bobroff chart**
60 million years	6′	8.8	1	C20/D27

It has a diameter of about 15 light years and was discovered in 1787 by William Herschel. It is located in the Perseus arm of our Galaxy, which is the next spiral arm outward from our position in the Galaxy.

NGC 457

Constellation	Object type	RA, Dec	Approx. transit date at local midnight	Distance
Cassiopeia	Open cluster	01h 19.5m, +58° 17′	October 20 (Standard Time)	8 thousand light years
Age	**Apparent size**	**Magnitude**	**Sky Atlas 2000.0 chart**	**Herald–Bobroff chart**
20 million years	13′	6.4	1	C20/D27

Nicknamed the "Owl Cluster" or the "E.T." cluster (the latter from its resemblance to the title character in the Spielberg movie). It has a diameter of about 30 light years and was discovered in 1787 by William Herschel. It is located in the Perseus arm of our Galaxy, which is the next spiral arm outward from our position in the Galaxy. The two bright stars on the SE edge of the cluster (φ Cas, mag. 5.0 and HD 7902, mag. 7.0) belong to the multiple star ADS 1073, which is a five star system and believed to be a member of this cluster.

NGC 488

Constellation	Object type	RA, Dec	Approx. transit date at local midnight	Distance
Pisces	Spiral galaxy	01h 21.8m, +05° 15′	October 21 (Standard Time)	1hundred million light years
Age	**Apparent size**	**Magnitude**	**Sky Atlas 2000.0 chart**	**Herald–Bobroff chart**
	5.4′ × 3.9′	10.3	10	C56

It has an unusually high rotational velocity (360 km/s at a distance of 65 thousand light years from the nucleus). The rotation axis is inclined at an angle of about 40° from our line-of-sight. It is also a tightly wound spiral, with the spiral arms having a pitch angle of only 5° (i.e. the angle between a spiral arm and a circle).

NGC 524

Constellation	Object type	RA, Dec	Approx. transit date at local midnight	Distance
Pisces	Lenticular galaxy (see M 102/ NGC 5866 for explanation)	01h 24.8m, +09° 32′	October 22 (Standard Time)	1 hundred million light years

Age	Apparent size	Magnitude	Sky Atlas 2000.0 chart	Herald–Bobroff chart
	3.2′ × 3.2′	10.3	10	C56

It has an optical diameter of about 90 thousand light years. It is nearly face-on but with a "warped" disk so that the inner (central) 20″ of the galaxy (which contains about 10 billion solar masses) lies in a plane that is skewed by about 20° from the rest of the galaxy. It is the namesake member of the NGC 524 galaxy group, which contains nine nearby (and gravitationally bound) galaxies, including NGC 489 (mag. 12.6, 49′ WSW) and the dimmer NGC 502 (mag. 12.8, 41′ SW), NGC 516 (mag. 13.1, 10′ W), NGC 518 (mag. 13.3, 14′ SSW), and NGC 532 (mag. 12.9, 17′ SSE).

NGC 559

Constellation	Object type	RA, Dec	Approx. transit date at local midnight	Distance
Cassiopeia	Open cluster	01h 29.5m, +63° 19′	October 23 (Standard Time)	4 thousand light years

Age	Apparent size	Magnitude	Sky Atlas 2000.0 chart	Herald–Bobroff chart
1 billion years	4.4′	9.5	1	C20/D27

It has a diameter of about 5 light years and was discovered in 1787 by William Herschel.

NGC 581 (M 103)

Constellation	Object type	RA, Dec	Approx. transit date at local midnight	Distance
Cassiopeia	Open cluster	01h 33.4m, +60° 39′	October 24 (Standard Time)	9 thousand light years

Age	Apparent size	Magnitude	Sky Atlas 2000.0 chart	Herald–Bobroff chart
20 million years	6.0′	7.4	1	C20/D27

Of 228 stars with magnitude 14.5 and brighter that have been examined in professional telescopic studies in the cluster region, 77 are known to be actual cluster members, the rest being field stars (see M 39/NGC 7092 for further discussion). The brightest star in the cluster region (a double star, Struve 131, mag. 7.3 and 10.5 with separation 13.8″ along a SE–NW direction) is not a cluster member, but is a field star (in the foreground). M 103's size of 6′ corresponds to a diameter of about 15 light years.

NGC 584

Constellation	Object type	RA, Dec	Approx. transit date at local midnight	Distance
Cetus	Elliptical galaxy	01h 31.3m, –06° 52′	October 24 (Standard Time)	80 million light years
Age	Apparent size	Magnitude	Sky Atlas 2000.0 chart	Herald–Bobroff chart
	2.8′ × 1.9′	10.5	10	C56

It has an unusually high rotational velocity for its galaxy type. It is the namesake member of the NGC 584 galaxy group (containing eight galaxies, three of which are elliptical and the others spirals) with NGC 586 (spiral galaxy, mag. 13.2 and 5′ ESE), NGC 596 (elliptical galaxy, mag. 10.9 and 26′ ESE – see NGC 596), NGC 600 (barred spiral galaxy, mag. 12.4, 38′ SSE), NGC 615 (spiral galaxy, 1° ESE – see NGC 615), NGC 636 (elliptical galaxy, mag. 11.3, 2° ESE) and two other dimmer galaxies, all of which are within a few million light years of each other.

NGC 596

Constellation	Object type	RA, Dec	Approx. transit date at local midnight	Distance
Cetus	Elliptical galaxy	01h 32.9m, –07° 02′	October 24 (Standard Time)	80 million light years
Age	Apparent size	Magnitude	Sky Atlas 2000.0 chart	Herald–Bobroff chart
	2.8′ × 2.1′	10.9	10	C56

In professional telescopic studies this galaxy has an odd internal structure that consists of several misaligned ellipsoidal shells. It is part of the NGC 584 galaxy group (see NGC 584).

NGC 598 (M 33)

Constellation	Object type	RA, Dec	Approx. transit date at local midnight	Distance
Triangulum	Spiral galaxy	01h 33.9m, +30° 39′	October 24 (Standard Time)	2.8 million light years
Age	Apparent size	Magnitude	Sky Atlas 2000.0 chart	Herald–Bobroff chart
	68.7′ × 41.6′	5.7	4	C38/D42

This is the third most luminous galaxy in our Local Group (which contains about 40 galaxies in a radius of a few million light years) after M 31 and the Milky Way. It is nicknamed the "Pinwheel Galaxy". It has an optical diameter of about 60 thousand light years (about half that of the Milky Way), which is about average for a spiral galaxy. Its luminous mass is about 10 billion Suns. It rotates clockwise from our viewpoint with a period of about 2 hundred million years. The size of its central black hole (if one exists) is at most about 3 thousand solar masses. NGC 604, the largest known ionized hydrogen region (1.5 thousand light years in diameter), belongs to this galaxy and is visible as a knot in large amateur telescopes near the NNE edge.

NGC 613

Constellation	Object type	RA, Dec	Approx. transit date at local midnight	Distance
Sculptor	Barred spiral galaxy	01h 34.3m, −29° 25′	October 24 (Standard Time)	70 million light years

Age	Apparent size	Magnitude	Sky Atlas 2000.0 chart	Herald–Bobroff chart
	5.2′ × 4.3′	10.1	18	C74

This is a Seyfert galaxy (where a supermassive object in this galaxy's center accumulates nearby gas resulting in strong emission from the nucleus). In addition to a large-scale bar, professional telescopic studies indicate this galaxy has a second bar structure in the inner nucleus (at radius < 6″), making this an unusual "double-barred" galaxy.

NGC 615

Constellation	Object type	RA, Dec	Approx. transit date at local midnight	Distance
Cetus	Spiral galaxy	01h 35.1m, −07° 20′	October 25 (Standard Time)	80 million light years

Age	Apparent size	Magnitude	Sky Atlas 2000.0 chart	Herald–Bobroff chart
	3.4′ × 1.4′	11.6	10	C56

It has an optical diameter of about 70 thousand light years. The rotation axis of its spiral disk is inclined at an angle of about 60° from our line-of-sight. The nucleus of this galaxy is decoupled from its outer regions (being both chemically and kinematically different), perhaps due to gravitational interaction with another galaxy some billions of years ago. It is part of the NGC 584 galaxy group containing eight galaxies (see NGC 584).

NGC 628 (M 74)

Constellation	Object type	RA, Dec	Approx. transit date at local midnight	Distance
Pisces	Spiral galaxy	01h 36.7 m, +15° 47′	October 25 (Standard Time)	30 million light years

Age	Apparent size	Magnitude	Sky Atlas 2000.0 chart	Herald–Bobroff chart
	10.0′ × 9.4′	9.4	10	C38

It has a mass of about 330 billion Suns and an optical diameter of almost 1 hundred thousand light years. It is nearly face-on (its inclination angle, which is the angle between its axis of rotation and our line-of-sight, is about 7°). Hundreds of ionized hydrogen (HII) star-forming regions (like M 42) have been identified in this galaxy in professional telescopes, some of which can be seen as bright knots in large amateur telescopes. All told, these star-forming regions produce about one new star per year. M 74 is part of a gravitationally bound group of seven galaxies that includes NGC 660 (2° 38′ SE) as well as five dimmer galaxies (the brightest of which, at mag. 13, are UGC 1195, 22′ NNW of NGC 660 and UGC 1200, 29′ S of NGC 660). Professional telescopic studies show that beyond its optically visible disk is an extended disk of atomic hydrogen reaching out more than twice its optical diameter. This extended disk is distorted for reasons that remain unknown, but may involve past interactions with other galaxies.

NGC 637

Constellation	Object type	RA, Dec	Approx. transit date at local midnight	Distance
Cassiopeia	Open cluster	01h 43.1m, +64° 02'	October 27 (Standard Time)	7 thousand light years
Age	**Apparent size**	**Magnitude**	**Sky Atlas 2000.0 chart**	**Herald–Bobroff chart**
10 million years	3.5'	8.2	1	C20/D27

It has a diameter of about 7 light years and was discovered in 1787 by William Herschel. It is located in the Perseus spiral arm of our Galaxy, which is the next arm outward from the Orion arm that we are located within.

NGC 650/651 (M 76)

Constellation	Object type	RA, Dec	Approx. transit date at local midnight	Distance
Perseus	Planetary nebula	01h 42.3m, +51° 34'	October 27 (Standard Time) ·	4 thousand light years
Age	**Apparent size**	**Magnitude**	**Sky Atlas 2000.0 chart**	**Herald–Bobroff chart**
	2'	11	1	C20

Nicknamed the "Little Dumbbell Nebula". Although different parts are expanding at different rates, typical expansion velocities are about 50 km/s. It has a diameter of about 2 light years. The two lobes of the "bowtie" shape are each three-dimensional expanding bubbles, inclined at 75° from our line-of-sight with the NW lobe pointing toward us.

NGC 654

Constellation	Object type	RA, Dec	Approx. transit date at local midnight	Distance
Cassiopeia	Open cluster	01h 44.0m, +61° 53'	October 27 (Standard Time)	7 thousand light years
Age	**Apparent size**	**Magnitude**	**Sky Atlas 2000.0 chart**	**Herald–Bobroff chart**
15 million years	5'	6.5	1	C20/D27

Although a young cluster, nearly all of its members have reached the main sequence. In professional telescopic studies it has more than 60 members (i.e. stars with > 50% probability of not being a field star). It has a diameter of about 10 light years.

NGC 659

Constellation	Object type	RA, Dec	Approx. transit date at local midnight	Distance
Cassiopeia	Open cluster	01h 44.4m, +60° 40'	October 27 (Standard Time)	6 thousand light years
Age	**Apparent size**	**Magnitude**	**Sky Atlas 2000.0 chart**	**Herald–Bobroff chart**
40 million years	5'	7.9	1	C20/D27

It has a diameter of nearly 10 light years and was discovered by Caroline Herschel (William Herschel's sister) in 1783. It contains at least seven variable stars.

NGC 663

Constellation	Object type	RA, Dec	Approx. transit date at local midnight	Distance
Cassiopeia	Open cluster	01h 46.0m, +61° 14′	October 27 (Standard Time)	6 thousand light years
Age	**Apparent size**	**Magnitude**	**Sky Atlas 2000.0 chart**	**Herald– Bobroff chart**
20 million years	16′	7.1	1	C20/D27

It has a diameter of about 30 light years. It has at least 15 variable stars as well as one of the largest percentages of "Be" stars known in any open cluster. Indeed, about 25% of its members are Be stars, which are B-type stars that are peculiar because of hydrogen Balmer emission lines in their spectra, due to material expelled by high rotational velocities into a circumstellar disk in the equatorial plane of the star – see M 47/NGC 2422.

NGC 720

Constellation	Object type	RA, Dec	Approx. transit date at local midnight	Distance
Cetus	Elliptical galaxy	01h 53.0m, −13° 44′	October 29 (Standard Time)	80 million light years
Age	**Apparent size**	**Magnitude**	**Sky Atlas 2000.0 chart**	**Herald– Bobroff chart**
	4.4′ × 2.2′	10.2	10	C56

It has a total mass of about a trillion Suns, although most of this is dark matter since the luminous stars have a mass that is less than 25% that of the dark matter mass in this galaxy. It contains nearly 7 hundred globular clusters.

NGC 752

Constellation	Object type	RA, Dec	Approx. transit date at local midnight	Distance
Andromeda	Open cluster	01h 57.7m, +37° 47′	October 30 (Standard Time)	1 thousand light years
Age	**Apparent size**	**Magnitude**	**Sky Atlas 2000.0 chart**	**Herald– Bobroff chart**
1 billion years	50′	5.7	4	C38/D41

It has about 135 members in professional telescopic studies. Despite being rather old (1 billion years), this cluster has retained much of its mass because it orbits at a large radius from the center of the Galaxy (no closer than 28 thousand light years). Open clusters that orbit closer in to the Galactic center are disrupted more readily, so that almost all old open clusters have orbits that stay at least as far from the Galactic center as we are located (i.e. at radii > 25 thousand light years or so).

NGC 772

Constellation	Object type	RA, Dec	Approx. transit date at local midnight	Distance
Aries	Spiral galaxy	01h 59.3m, +19° 00'	October 31 (Standard Time)	110 million light years
Age	**Apparent size**	**Magnitude**	**Sky Atlas 2000.0 chart**	**Herald–Bobroff chart**
	7.5' × 4.3'	10.3	10	C38

It rotates about an axis that is inclined at 54° from our line-of-sight. It has two faint dwarf companion galaxies (UGC 1519, mag. 16, 43' ENE, and UGC 1546, mag. 15, 1.0° ESE), as well as a satellite galaxy (NGC 770, mag. 12.9 and 4' SSW), the latter rotating around NGC 772 in a retrograde orbit (i.e. opposite to the direction of rotation of NGC 772).

NGC 779

Constellation	Object type	RA, Dec	Approx. transit date at local midnight	Distance
Cetus	Barred spiral galaxy	01h 59.7m, –05° 58'	October 31 (Standard Time)	60 million light years
Age	**Apparent size**	**Magnitude**	**Sky Atlas 2000.0 chart**	**Herald–Bobroff chart**
	4.1' × 1.2'	11.2	10	C56

It has an optical diameter of about 70 thousand light years and a mass of about 50 billion Suns. It is nearly edge-on (inclination angle of about 75°, meaning it rotates about an axis that is inclined at 75° from our line-of-sight).

NGC 869

Constellation	Object type	RA, Dec	Approx. transit date at local midnight	Distance
Perseus	Open cluster	02h 19.1m, +57° 08'	November 5	7 thousand light years
Age	**Apparent size**	**Magnitude**	**Sky Atlas 2000.0 chart**	**Herald–Bobroff chart**
10 million years	30'	5.3	1	C19/D27

Also known as *h* Persei. Combined with NGC 884 it forms the "Double Cluster" – see NGC 884. The two clusters are thought to be at nearly the same distance and have approximately the same age. NGC 869 has a diameter of about 60 light years and a mass of about 4 thousand suns. Professional telescopic studies indicate it contains several hundred members.

NGC 884

Constellation	Object type	RA, Dec	Approx. transit date at local midnight	Distance
Perseus	Open cluster	02h 22.5m, +57° 09′	November 5	7 thousand light years
Age	**Apparent size**	**Magnitude**	**Sky Atlas 2000.0 chart**	**Herald–Bobroff chart**
10 million years	30′	6.1	1	C19/D27

Also known as χ Persei. It has a diameter of about 60 light years and a mass of about 3 thousand Suns (i.e. it is about ¾ as massive as NGC 869). Together with NGC 869 it forms the "Double Cluster" – see NGC 869. Together the two clusters contain 33 known Be stars, which is a much higher fraction of Be stars than in the surrounding field stars (see NGC 663 for a definition of Be stars). This is thought to be related to an enhancement in the proportion of Be stars that occurs as stars approach the end of their main-sequence lives. Certain stars at this evolutionary stage expel material into a circumstellar disk, for reasons that remain a topic of debate. Since the cluster has proportionately more stars at this evolutionary stage than in the field stars, it also has proportionately more Be stars than in the surrounding field. Professional telescopic studies indicate it contains several hundred stars.

NGC 891

Constellation	Object type	RA, Dec	Approx. transit date at local midnight	Distance
Andromeda	Spiral galaxy	02h 22.6m, +42° 21′	November 5	30 million light years
Age	**Apparent size**	**Magnitude**	**Sky Atlas 2000.0 chart**	**Herald–Bobroff chart**
	13.1′ × 2.8′	9.9	4	C37/D41

It has an optical diameter of about 120 thousand light years. The dust lane visible in large amateur telescopes has a width of about 1.5 thousand light years (10″). Professional telescopic studies indicate that the Galaxy is not perfectly edge-on, but instead the E side is inclined just slightly toward us and the W side just slightly away from us (the rotation axis is inclined at an angle of 89° from our line-of-sight; 90° would be exactly edge-on). It is part of the NGC 1023 group of eight galaxies (see NGC 1023). This galaxy is thought to be quite similar to what our Galaxy would look like if viewed edge-on.

NGC 908

Constellation	Object type	RA, Dec	Approx. transit date at local midnight	Distance
Cetus	Barred spiral galaxy	02h 23.1m, −21° 14′	November 6	60 million light years
Age	**Apparent size**	**Magnitude**	**Sky Atlas 2000.0 chart**	**Herald–Bobroff chart**
	6.1′ × 2.7′	10.2	18	C73

It has an optical diameter of 110 thousand light years. It is the largest member of the NGC 908 group of galaxies, which contains four other spirals and four irregular galaxies, including NGC 907 (barred spiral, mag. 12.6, 31′ N), IC 223 (irregular galaxy, mag. 13, 33′ NNW), and NGC 899 (irregular galaxy, mag. 12.6, 31′ NNW).

NGC 936

Constellation	Object type	RA, Dec	Approx. transit date at local midnight	Distance
Cetus	Barred lenticular galaxy	02h 27.6m, –01° 09′	November 7	55 million light years

Age	Apparent size	Magnitude	Sky Atlas 2000.0 chart	Herald–Bobroff chart
	4.3′ × 3.8′	10.1	10	C55

It has an optical diameter of about 70 thousand light years. It is the brightest member of the NGC 936 galaxy group that includes seven galaxies, including NGC 941 (barred spiral, mag. 12.4, 13′ E) and NGC 955 (spiral, mag. 12.0, 45′ E). Professional telescopic studies show its bar extends to an apparent distance of 41″ and rotates as a solid body with a period of about one hundred million years, rotating about an axis that is inclined 41° to our line-of-sight.

NGC 1022

Constellation	Object type	RA, Dec	Approx. transit date at local midnight	Distance
Cetus	Barred spiral galaxy	02h 38.5m, –06° 41′	November 9	60 million light years

Age	Apparent size	Magnitude	Sky Atlas 2000.0 chart	Herald–Bobroff chart
	2.6′ × 2.5′	11.3	10	C55

It rotates about an axis inclined at about 25° from our line-of-sight. It is part of the NGC 1052 group (see NGC 1052). It has an optical diameter of about 50 thousand light years.

NGC 1023

Constellation	Object type	RA, Dec	Approx. transit date at local midnight	Distance
Perseus	Barred lenticular galaxy	02h 40.4m, +39° 04′	November 10	30 million light years

Age	Apparent size	Magnitude	Sky Atlas 2000.0 chart	Herald–Bobroff chart
	8.1′ × 3.4′	9.4	4	C37/D41

The center of this galaxy is thought to contain a supermassive black hole that has a mass about 40 million times that of our Sun. Stars near this black hole are moving at very high velocities (> 6 hundred km/s), and the visible nucleus is believed to be disk shaped. The diffuse extension off the eastern end, visible in large amateur telescopes, is a dwarf galaxy (NGC 1023A) that is in the process of being accreted into NGC 1023. This merger is probably one of several such mergers in NGC 1023's past, and more mergers are likely since NGC 1023 is the namesake member of the gravitationally bound NGC 1023 group of galaxies, which also contains NGC 1003 (spiral galaxy, 1° 48′ N, mag. 11.4), IC 239 (barred spiral galaxy, 46′ W, mag. 10) plus at least two other smaller galaxies.

NGC 1027

Constellation	Object type	RA, Dec	Approx. transit date at local midnight	Distance
Cassiopeia	Open cluster	02h 42.6m, +61° 36'	November 10	3 thousand light years
Age	Apparent size	Magnitude	Sky Atlas 2000.0 chart	Herald–Bobroff chart
160 million years	20'	6.7	1	C19/D27

It has a diameter of about 20 light years and was discovered in 1786 by William Herschel.

NGC 1039 (M 34)

Constellation	Object type	RA, Dec	Approx. transit date at local midnight	Distance
Perseus	Open cluster	02h 42.1m, +42° 46'	November 10	1.5 thousand light years
Age	Apparent size	Magnitude	Sky Atlas 2000.0 chart	Herald–Bobroff chart
250 million years	35'	5.2	4	C37/D41

It has a diameter of about 15 light years and was discovered in the middle of the seventeenth century by Giovanni Batista Hodierna. Its stars rotate at rates that are midway between those in the younger Pleiades cluster (100 million years old – see M 45) and the older Hyades cluster (6 hundred million years old). This is thought to be the result of rotational braking whose effect on rotation rates becomes more pronounced with age. Such braking is believed to be due to angular momentum loss via magnetic coupling to the chromosphere (i.e. the star's atmosphere outside the bright photosphere).

NGC 1052

Constellation	Object type	RA, Dec	Approx. transit date at local midnight	Distance
Cetus	Elliptical galaxy	02h 41.1m, −08° 15'	November 10	60 million light years
Age	Apparent size	Magnitude	Sky Atlas 2000.0 chart	Herald–Bobroff chart
	2.8' × 2.0'	10.5	10	C55

It is believed to contain a supermassive central black hole with a mass of about one hundred million Suns. This black hole is believed to be responsible for low-ionization nuclear emission regions (LINERs – see M 81/NGC 3031 for an explanation) and two high-speed jets (traveling at 50% the speed of light) directed oppositely and emanating from the core. It contains globular clusters of two quite different ages which may have belonged to two distinct galaxies that previously merged to form NGC 1052. It is the namesake member of the NGC 1052 group of galaxies, consisting of 14 gravitationally bound galaxies including NGC 988 (barred spiral, mag. 11, 1° 47' SW), NGC 1022 (1° 42' NNW – see NGC 1022), NGC 1035 (spiral, mag. 12.2, 25' WNW), NGC 1042 (barred spiral, mag. 10.9, 15' SW), NGC 1084 (spiral, mag. 10.6, 1° 23' ENE) and NGC 1140 (irregular galaxy, mag. 12.5, 3° 47' ESE).

NGC 1055

Constellation	Object type	RA, Dec	Approx. transit date at local midnight	Distance
Cetus	Barred spiral galaxy	02h 41.8m, +00° 27'	November 10	40 million light years
Age	**Apparent size**	**Magnitude**	**Sky Atlas 2000.0 chart**	**Herald–Bobroff chart**
	7.6' × 3.0'	10.6	10	C55

This and nearby M 77 (30' SSE – see NGC 1068/M 77) are both part of a galaxy group of six gravitationally bound galaxies that also includes NGC 1073 (barred spiral, mag. 11.0, 1° NNE), plus other non-NGC galaxies. M 77's projected distance (i.e. distance perpendicular to our line-of-sight) from NGC 1055 is about 4 hundred thousand light years.

NGC 1068 (M 77)

Constellation	Object type	RA, Dec	Approx. transit date at local midnight	Distance
Cetus	Spiral galaxy	02h 42.7m, –00° 01'	November 11	50 million light years
Age	**Apparent size**	**Magnitude**	**Sky Atlas 2000.0 chart**	**Herald–Bobroff chart**
	7.3' × 6.3'	8.9	10	C55

This is a well-known Seyfert galaxy, in which emission from the nucleus is thought to occur due to accretion of matter onto a massive central black hole (which is thought to have a mass of about 20 million Suns in this galaxy). It is part of a gravitationally bound group of six galaxies that includes NGC 1055 (30' NNW), NGC 1073 (1° 25' N), as well as the much dimmer, mag. 13, UGC 2275 and UGC 2302. M 77 also contains water-vapor "masers" in its central region. Maser stands for "microwave amplification by stimulated emission of radiation", the physics of which is the microwave equivalent of a laser, except that lasers are usually designed to produce a beam, while astronomical masers yield emission that radiates from a roughly spherical region. M 77 has an optical diameter of about 110 thousand light years.

NGC 1084

Constellation	Object type	RA, Dec	Approx. transit date at local midnight	Distance
Eridanus	Spiral galaxy	02h 46.0m, –07° 35'	November 11	60 million light years
Age	**Apparent size**	**Magnitude**	**Sky Atlas 2000.0 chart**	**Herald–Bobroff chart**
	3.5' × 2.1'	10.7	10	C55

It is part of the NGC 1052 galaxy group (see NGC 1052). It has an optical diameter of about 60 thousand light years. It contains a giant star-forming region in its NE edge. The Galaxy rotates about an axis that is inclined 57° to our line-of-sight, with the SE edge lying closest to us. It is one of fewer than ten galaxies that has had four or more recorded supernovae.

NGC 1232

Constellation	Object type	RA, Dec	Approx. transit date at local midnight	Distance
Eridanus	Barred spiral galaxy	03h 09.8m, −20° 35′	November 17	70 million light years
Age	**Apparent size**	**Magnitude**	**Sky Atlas 2000.0 chart**	**Herald–Bobroff chart**
	7.1′ × 6.3′	9.9	18	C73/D39

This galaxy rotates about an axis that is inclined from our line-of-sight by about 30°, meaning that its disk is nearly face-on to us. In professional telescopes it is seen to have a multiple arm nature instead of the more usual two spiral arms. Ionized hydrogen (HII) star-forming regions are present throughout its arms, collections of which can be seen as brighter "knots" in large amateur telescopes. Having a mass of nearly 1/2 trillion Suns, this is a massive galaxy (although about 2/3 of this is dark matter). Its optical diameter is also large at nearly 150 thousand light years.

NGC 1245

Constellation	Object type	RA, Dec	Approx. transit date at local midnight	Distance
Perseus	Open cluster	03h 14.7m, +47° 14′	November 19	8 thousand light years
Age	**Apparent size**	**Magnitude**	**Sky Atlas 2000.0 chart**	**Herald–Bobroff chart**
1 billion years	10′	8.4	4	C19

It lies 35 thousand light years from the center of our Galaxy (about 10 thousand light years farther out than us) and about 1.3 thousand light years below the Galactic plane. Although previously thought to be abnormally rich in heavier elements, recent data suggest this is not the case.

NGC 1342

Constellation	Object type	RA, Dec	Approx. transit date at local midnight	Distance
Perseus	Open cluster	03h 31.7m, +37° 22′	November 23	2 thousand light years
Age	**Apparent size**	**Magnitude**	**Sky Atlas 2000.0 chart**	**Herald–Bobroff chart**
4 hundred million years	14′	6.7	4	C36

It has a diameter of about 8 light years and was discovered in 1799 by William Herschel. Professional telescopic studies to date count about 120 member stars.

NGC 1407

Constellation	Object type	RA, Dec	Approx. transit date at local midnight	Distance
Eridanus	Elliptical galaxy	03h 40.2m, −18° 35′	November 25	60 million light years
Age	**Apparent size**	**Magnitude**	**Sky Atlas 2000.0 chart**	**Herald–Bobroff chart**
	4.9′ × 4.5′	9.7	11	C72/D39

This galaxy is part of the Eridanus A group of galaxies which contains about 50 gravitationally bound galaxies and includes nearby NGC 1400 (elliptical galaxy, mag. 11.0, 12′ WSW). This group has an abnormally large amount of dark matter for reasons unknown. NGC 1407 and NGC 1400 emit about 80% of the total light from this galaxy group. For reasons that remain unexplained, NGC 1400 has a much lower velocity away from us (549 km/s) than the rest of the galaxies in this cluster (receding at an average of 1666 km/s).

NGC 1444

Constellation	Object type	RA, Dec	Approx. transit date at local midnight	Distance
Perseus	Open cluster	03h 49.4m, +52° 40′	November 27	4 thousand light years
Age	**Apparent size**	**Magnitude**	**Sky Atlas 2000.0 chart**	**Herald–Bobroff chart**
1 hundred million years	4′	6.6	1	C18

It is located 30 thousand light years from the Galactic center and less than one hundred light years below the Galactic center plane. It has a diameter of about 5 light years.

NGC 1491

Constellation	Object type	RA, Dec	Approx. transit date at local midnight	Distance
Perseus	Emission nebula	04h 03.2m, +51° 19′	December 1	12 thousand light years
Age	**Apparent size**	**Magnitude**	**Sky Atlas 2000.0 chart**	**Herald–Bobroff chart**
	6′ × 9′	8.5	1	C18

This nebula is an ionized hydrogen (HII) region that is part of a larger molecular gas cloud (which is not visible in amateur telescopes) that has a mass of tens of thousands of Suns. The part we are seeing is made visible by ionizing radiation from the young, hot, mag 11.2 star (BD +50 886), which is the brightest star in the immediate vicinity and is just east of the brightest portions of the nebula. The star appears separated from the nebula because its stellar winds have cleared out a hemispherical cavity in the gas cloud adjacent to (west of) the star.

NGC 1501

Constellation	Object type	RA, Dec	Approx. transit date at local midnight	Distance
Camelopardalis	Planetary nebula	04h 07.0m, +60° 55′	December 2	5 thousand light years

Age	Apparent size	Magnitude	Sky Atlas 2000.0 chart	Herald–Bobroff chart
	52″	12	1	C18

It is classified as an "elliptical" planetary nebula. The central star (mag. 14.5) is a Wolf–Rayet star (see NGC 40 for an explanation) with a mass of about 0.55 Suns. The nebula has a diameter of about one light year. It is expanding outward at about 80 km/s.

NGC 1502

Constellation	Object type	RA, Dec	Approx. transit date at local midnight	Distance
Camelopardalis	Open cluster	04h 07.8m, +62° 20′	December 2	3 thousand light years

Age	Apparent size	Magnitude	Sky Atlas 2000.0 chart	Herald–Bobroff chart
10 million years	8′	6.9	1	C18

It has a diameter of about 6 light years. The two brightest members of this cluster, near the center of the cluster, are a visual binary pair (separation 18″). The more northern of the two stars itself is a quadruple system (which includes an eclipsing binary pair) and the more southern star is a spectroscopic binary, so that this is actually a six star system (despite its binary appearance in amateur telescopes). The cluster lies at the SE end of "Kemble's Cascade" (a pretty, 2.5° long line of stars visible in binoculars and named after the Canadian amateur astronomer Lucien Kemble).

NGC 1513

Constellation	Object type	RA, Dec	Approx. transit date at local midnight	Distance
Perseus	Open cluster	04h 09.9m, +49° 31′	December 3	4 thousand light years

Age	Apparent size	Magnitude	Sky Atlas 2000.0 chart	Herald–Bobroff chart
150 million years	9′	8.4	5	C18

It has a diameter of about 10 light years. It lies 30 thousand light years from the center of our Galaxy (about 5 thousand light years farther out than us) and 70 light years below the Galactic central plane. It was discovered in 1790 by William Herschel.

NGC 1514

Constellation	Object type	RA, Dec	Approx. transit date at local midnight	Distance
Taurus	Planetary nebula	04h 09.3m, +30° 47′	December 3	4 thousand light years
Age	Apparent size	Magnitude	Sky Atlas 2000.0 chart	Herald–Bobroff chart
	2′	11	5	C36

The central star (mag. 9.4) is evident in amateur telescopes and known to be a spectroscopic binary. Like many planetary nebulae, it has a multiple shell nature. The part made visible in amateur telescopes by doubly ionized oxygen (OIII) is expanding outward at about 50 km/s.

NGC 1528

Constellation	Object type	RA, Dec	Approx. transit date at local midnight	Distance
Perseus	Open cluster	04h 15.3m, +51° 13′	December 4	3 thousand light years
Age	Apparent size	Magnitude	Sky Atlas 2000.0 chart	Herald–Bobroff chart
4 hundred million years	24′	6.4	1	C18

It has a diameter of nearly 20 light years. It was discovered in 1790 by William Herschel. It contains about 170 member stars with membership probability > 50% in professional telescopic studies.

NGC 1535

Constellation	Object type	RA, Dec	Approx. transit date at local midnight	Distance
Eridanus	Planetary nebula	04h 14.3m, −12° 44′	December 4	6 thousand light years
Age	Apparent size	Magnitude	Sky Atlas 2000.0 chart	Herald–Bobroff chart
	20″	10	11	C54

Turbulence causes fluctuating velocities of about 8 km/s in this nebula, which is thousands of times the fluctuating velocities of turbulent eddies on a windy day on Earth. Professional telescopic studies show an inner shell (diameter near 20″) expanding at tens of km/s that is constricted by a torus of material, in addition to a dim outer shell (with a diameter of nearly 50″).

NGC 1545

Constellation	Object type	RA, Dec	Approx. transit date at local midnight	Distance
Perseus	Open cluster	04h 20.9m, +50° 15′	December 5	2 thousand light years
Age	Apparent size	Magnitude	Sky Atlas 2000.0 chart	Herald–Bobroff chart
3 hundred million years	18′	6.2	1	C18

It lies 30 thousand light years out from the center of the Galaxy, but lies within a few light years of the central plane of our Galaxy.

NGC 1647

Constellation	Object type	RA, Dec	Approx. transit date at local midnight	Distance
Taurus	Open cluster	04h 45.9m, +19° 06′	December 12	2 thousand light years
Age	**Apparent size**	**Magnitude**	**Sky Atlas 2000.0 chart**	**Herald–Bobroff chart**
150 million years	45′	6.4	11	C35

It has a diameter of a little over 20 light years. It contains nearly 2 hundred stars in professional telescopic studies. Despite lying 2° to the WSW of the cluster center, the Cepheid variable star SZ Tau (mag. 6.5) is thought to be an outlying member of this cluster since it shares a similar proper motion, radial velocity and age as the cluster (see NGC 7790 for a brief explanation of Cepheid variables). SZ Tau has a radius about 35 times that of our Sun.

NGC 1664

Constellation	Object type	RA, Dec	Approx. transit date at local midnight	Distance
Auriga	Open cluster	04h 51.1m, +43° 41′	December 13	4 thousand light years
Age	**Apparent size**	**Magnitude**	**Sky Atlas 2000.0 chart**	**Herald–Bobroff chart**
3 hundred million years	18′	7.6	5	C35/D25

It has a diameter of about 20 light years. It contains nearly 150 member stars in professional telescopic studies. Stars near the center of this cluster are spaced an average of about 1/2 a light year apart. Interstellar distances in the cluster increase for stars farther from the center, so that near the cluster border, interstellar spacings are near those in the neighborhood of our Sun (where the interstellar spacing is about 10 light years).

NGC 1788

Constellation	Object type	RA, Dec	Approx. transit date at local midnight	Distance
Orion	Reflection nebula	05h 06.9m, −03° 20′	December 17	2 thousand light years
Age	**Apparent size**	**Magnitude**	**Sky Atlas 2000.0 chart**	**Herald–Bobroff chart**
	2′		11	C53/D38

It has a diameter of about 1 light year. A young star cluster (less than a few million years old) is thought to be forming in the vicinity of this nebula.

NGC 1817

Constellation	Object type	RA, Dec	Approx. transit date at local midnight	Distance
Taurus	Open cluster	05h 12.2m, +16° 41′	December 18	6 thousand light years
Age	**Apparent size**	**Magnitude**	**Sky Atlas 2000.0 chart**	**Herald–Bobroff chart**
4 hundred million years	16′	7.7	11	C35

It has a diameter of nearly 30 light years. It contains almost 6 hundred member stars with membership probability > 50% in professional telescopic studies. It is located about 6 thousand light years almost directly outward from the Galactic center from us and 1 thousand light years below the Galactic plane. The open cluster NGC 1807 is 22′ WSW.

NGC 1857

Constellation	Object type	RA, Dec	Approx. transit date at local midnight	Distance
Auriga	Open cluster	05h 19.1m, +39° 21′	December 20	6 thousand light years
Age	**Apparent size**	**Magnitude**	**Sky Atlas 2000.0 chart**	**Herald–Bobroff chart**
2 hundred million years	6′	7.0	5	C35/D25

It has a diameter of a little over 10 light years. It lies directly outward from the Galactic center from us, close to the Galactic central plane (lying a little over one hundred light years below it). Although the open cluster Czernik 20 is very nearby (7′ N), Czernik 20 is in fact much farther away from us (nearly 11 thousand light years away) and is much younger than NGC 1857 (Czernik 20 is about 15 million years old).

NGC 1904 (M 79)

Constellation	Object type	RA, Dec	Approx. transit date at local midnight	Distance
Lepus	Globular cluster	05h 24.2m, −24° 31′	December 22	40 thousand light years
Age	**Apparent size**	**Magnitude**	**Sky Atlas 2000.0 chart**	**Herald–Bobroff chart**
12–14 billion years	7.8′	7.7	19	C71

It has a mass of almost 4 hundred thousand Suns and a diameter of about one hundred light years. Like most globular clusters it does not orbit our Galaxy with the Galactic disk as we do. Instead it follows an orbit that takes it out as far as 65 thousand light years from the Galactic center and then back in as close as about 14 thousand light years, on a path inclined to the Galactic disk (by about 30°), taking about 4 hundred million years to make one revolution around our Galaxy.

NGC 1907

Constellation	Object type	RA, Dec	Approx. transit date at local midnight	Distance
Auriga	Open cluster	05h 28.1m, +35° 20′	December 22	5 thousand light years
Age	**Apparent size**	**Magnitude**	**Sky Atlas 2000.0 chart**	**Herald–Bobroff chart**
4 hundred million years	7′	8.2	5	C35/D25

It has a diameter of about 10 light years. It is located about 6 thousand light years almost directly outward from the Galactic center from us and only 20 light years below the Galactic plane. M 38 (NGC 1912), 32′ N, has a projected distance (i.e. distance perpendicular to our line-of-sight) of about 40 light years from NGC 1907. It has been proposed that the two clusters formed from the same cloud since they share a reasonably similar age, similar amounts of elements heavier than helium, and similar velocities relative to us, making them a double cluster if proven true.

NGC 1912 (M 38)

Constellation	Object type	RA, Dec	Approx. transit date at local midnight	Distance
Auriga	Open cluster	05h 28.7m, +35° 51′	December 23	4 thousand light years
Age	**Apparent size**	**Magnitude**	**Sky Atlas 2000.0 chart**	**Herald–Bobroff chart**
3 hundred million years	21′	6.4	5	C35/D25

It has a diameter of about 25 light years and lies about 50 light years above the Galactic central plane. It is physically close to M 37 and M 36 (see M 36). NGC 1907, 32′ SSW, has a projected distance (i.e. the distance perpendicular to our line-of-sight) of about 40 light years from M 38. It has been proposed that the two clusters formed from the same cloud (since they share a reasonably similar age, similar amounts of elements heavier than helium, and similar velocities relative to us), making them a double cluster if proven true.

NGC 1931

Constellation	Object type	RA, Dec	Approx. transit date at local midnight	Distance
Auriga	Open cluster + emission nebula	05h 31.4m, +34° 15′	December 23	7 thousand light years
Age	**Apparent size**	**Magnitude**	**Sky Atlas 2000.0 chart**	**Herald–Bobroff chart**
10 million years	3′	10.1	5	C35/D25

It has a diameter of about 6 light years. Because of its young age, the remains of the nebula from which the open cluster formed is prominent as the surrounding emission nebula. Fewer than 10 of the embedded open cluster members have magnitudes < 14.0 so that its "cluster" nature is not strongly apparent in amateur telescopes.

NGC 1952 (M 1)

Constellation	Object type	RA, Dec	Approx. transit date at local midnight	Distance
Taurus	Supernova remnant	05h 34.5m, +22° 01'	December 24	6.5 thousand light years
Age	**Apparent size**	**Magnitude**	**Sky Atlas 2000.0 chart**	**Herald–Bobroff chart**
1 thousand years (created by a supernova in 1054 AD)	8' × 4'	8.4	5	C35

Nicknamed the "Crab Nebula", this is one of the most well-studied objects in the sky. At its center is a pulsar, which is a rotating neutron star (about 10–15 km in diameter with a mass somewhat greater than the Sun) with a strong magnetic field that emits a narrow beam of radio emission. The Crab pulsar rotates about 30 times a second (i.e. its period is 33 milliseconds), with its beacon pointing at us once each rotation (like a very rapidly rotating lighthouse beam). Of the more than 1 thousand known radio pulsars, this is one of only five whose pulses are visible in optical wavelengths (with professional telescopes). Pulsars are the remains of a star that went supernova. For the Crab pulsar, the progenitor star is thought to have had a mass of 8–13 times that of our Sun (its exact mass is uncertain). The Crab Nebula constitutes the remnants of the material ejected by the supernova event of this star, with about five solar masses worth of material being present in the luminous portion of the nebula. The filaments in the nebula are moving outward at over a thousand km/s.

NGC 1960 (M 36)

Constellation	Object type	RA, Dec	Approx. transit date at local midnight	Distance
Auriga	Open cluster	05h 36.3m, +34° 08'	December 25	4 thousand light years
Age	**Apparent size**	**Magnitude**	**Sky Atlas 2000.0 chart**	**Herald–Bobroff chart**
20 million years	12'	6.0	5	C35/D25

It lies about 70 light years above the Galactic central plane and has a diameter of about 15 light years. Its neighbors M 37 (3° 45' ESE) and M 38 (2° 16' NW) lie within a thousand light years of M 36. Although it is quite a young cluster, it is old enough that almost all its stars have had time to lose their youthful circumstellar disks (that formed when the stars collapsed from the surrounding gas and dust – see NGC 2362).

NGC 1961

Constellation	Object type	RA, Dec	Approx. transit date at local midnight	Distance
Camelopardalis	Barred spiral galaxy	05h 42.1m, +69° 23'	December 26	180 million light years
Age	**Apparent size**	**Magnitude**	**Sky Atlas 2000.0 chart**	**Herald–Bobroff chart**
	4.5' × 3.1'	11	1	C17

This is one of the most massive and largest spiral galaxies known, having a dynamical mass as high as perhaps 10 trillion Suns and an optical diameter of 240 thousand light years (which is about 4 times the diameter of the average galaxy). Compare this to our Galaxy, which is considered a large spiral in its own right, having a mass of 3/4 trillion Suns and an optical diameter of about 1 hundred thousand light years.

NGC 1964

Constellation	Object type	RA, Dec	Approx. transit date at local midnight	Distance
Lepus	Barred spiral galaxy	05h 33.4m, −21° 57′	December 24	65 million light years
Age	**Apparent size**	**Magnitude**	**Sky Atlas 2000.0 chart**	**Herald–Bobroff chart**
	5.5′ × 2.1′	10.8	19	C71

A foreground star 3″ E of the nucleus is partly responsible for the starlike appearance of the nucleus. This galaxy has an optical diameter of about 1 hundred thousand light years and a mass of about 150 billion Suns.

NGC 1973/1975/1977

Constellation	Object type	RA, Dec	Approx. transit date at local midnight	Distance
Orion	Reflection and emission nebulae + open cluster	05h 35.1m, −04° 44′	December 24	1.6 thousand light years
Age	**Apparent size**	**Magnitude**	**Sky Atlas 2000.0 chart**	**Herald–Bobroff chart**
	20′ × 10′	7	11	C53/D24

These nebulae are merely the fluorescing parts of a single larger molecular cloud (which is not visible in amateur telescopes). Indeed, these three nebulae are all part of the group of giant molecular gas clouds called the Orion–Monoceros complex that M 42 also belongs to (see M 42). An embedded young open cluster (most of whose stars are obscured by the dust and gas in the nebulae) supplies the photons that make these nebulae visible (mostly by scattering off dust in the nebulae, but also by ionizing gas in the nebulae, resulting in emission). The bright star 42 Orionis (mag. 4.7) is responsible for most of the illuminating (although at the south end, some of the light is actually light scattered from M 42).

NGC 1976 (M 42)

Constellation	Object type	RA, Dec	Approx. transit date at local midnight	Distance
Orion	Emission and reflection nebula	05h 35.3m, −05° 23′	December 24	1.5 thousand light years
Age	**Apparent size**	**Magnitude**	**Sky Atlas 2000.0 chart**	**Herald–Bobroff chart**
	1.5° × 1°	4.0	11	C53/D24

Nicknamed the "Orion Nebula". This is the apparently brightest and one of the closest ionized hydrogen star-forming regions (or HII regions) in our sky. It is lit up by the stars in a very young cluster (only a few hundred thousand years old) situated in the heart of the nebula. However, many of the stars in this cluster are obscured by material in the nebula. Light from these stars is both scattered off dust, particularly in the outer regions of the nebula, and re-emitted by gas, particularly in the inner regions, making this both a reflection and emission nebula. Four stars in the cluster form a quadrangle (with sides of about 10–20″) called the "Trapezium" and are all part of the multiple star system θ^1 Orionis. Professional telescopic studies indicate θ^1 Orionis contains 14 stars, only six of which can be seen, including the four in the Trapezium, under good seeing conditions in moderate amateur telescopes. θ^1 Orionis is in fact a wide double star with θ^2 Orionis (itself a triple star, so that θ Orionis consists of 17 stars!). The brightest (most southern) star in the Trapezium quadrangle (θ^1 Orionis C) is responsible for the ionization of the nebula. M 42 is actually only a small "blister" on the near side of a much larger cloud of gas and dust (the Orion A complex). The Orion A complex is itself part of an even larger group of giant molecular gas clouds (the Orion–Monoceros complex) that extend 30° in a SE–NW direction and sit about 5 hundred light years below the Galactic central plane. The Orion–Monoceros complex may have formed as the result of the infall of a gargantuan cloud onto the Galactic disk from below. The 1.5° apparent dimension of M 42 corresponds to a diameter of 40 light years.

NGC 1980

Constellation	Object type	RA, Dec	Approx. transit date at local midnight	Distance
Orion	Emission nebula + open cluster	05h 35.4m, −05° 55′	December 24	1 thousand light years
Age	**Apparent size**	**Magnitude**	**Sky Atlas 2000.0 chart**	**Herald–Bobroff chart**
	14′	2.5	11	C53/D24

It sits on the southern edge of M 42 (the Orion Nebula) and has Iota Orionis (44 Orionis) embedded within it. Iota Orionis (mag. 2.8) is actually a spectroscopic binary that when paired with the mag. 7.7 star 11″ to the SE makes up a triple star system. It is speculated that the two stars making up Iota Orionis may each have been paired with a different star previously, but a close encounter approximately 2.5 million years ago resulted in the previous partner stars being flung away, leaving Iota in its current spectroscopic binary state. The two run-away stars that are thought to once have been paired with one of each star in Iota Orionis are AE Aurigae and μ Columbae.

NGC 1982 (M 43)

Constellation	Object type	RA, Dec	Approx. transit date at local midnight	Distance
Orion	Emission nebula	05h 35.5m, –05° 16′	December 24	1.5 thousand light years
Age	Apparent size	Magnitude	Sky Atlas 2000.0 chart	Herald–Bobroff chart
	20′	9.0	11	C53/D24

This is part of the same gas and dust cloud as M 42 (the Orion A complex), lying just NE of M 42, separated from M 42 by a dust lane between the two. M 43 has relatively little dust, so its light is largely from gas emission (it is an ionized hydrogen i.e. HII, region). The nebula is visible because it is ionized by the bright variable star NU Orionis (i.e. HD 37061, mag. 7) in its center. Professional telescopic studies find planet-forming disks ("proplyds") are present around several stars in M 43 (as well as in M 42).

NGC 1999

Constellation	Object type	RA, Dec	Approx. transit date at local midnight	Distance
Orion	Emission and reflection nebula	05h 36.4m, –06° 43′	December 25	1 thousand light years
Age	Apparent size	Magnitude	Sky Atlas 2000.0 chart	Herald–Bobroff chart
	2′		11	C53/D24

The nebula is illuminated by the star V380 (mag. 10.3), which is the only bright star near the center of the nebula. Professional telescopes find two Herbig–Haro objects nearby (HH 1, 3′ SW and HH 2 lying 6′ S). Herbig–Haro objects are thought to arise when surrounding gas and dust spiral into a young star. A portion of this accreting material is flung out perpendicular to the accretion disk in two opposing jets moving at a few hundred kilometers per hour. As the jets slam into the surrounding gas, shock waves arise that heat this gas into emission. The appearance of the resulting glowing jet in professional telescopic images has been likened to that of a flame-thrower.

NGC 2022

Constellation	Object type	RA, Dec	Approx. transit date at local midnight	Distance
Orion	Planetary nebula	05h 42.1m, +09° 05′	December 26	6 thousand light years
Age	Apparent size	Magnitude	Sky Atlas 2000.0 chart	Herald–Bobroff chart
	19″	12	11	C53

This is a prolate spheroid in shape (an extreme example of a prolate spheroid is a cigar), with a major to minor axis ratio of 1.2. Professional telescopes show a second, fainter, outer spherical shell expanding more slowly than the inner one seen in amateur telescopes. NGC 2022 has a diameter of about 1/2 light year.

NGC 2024

Constellation	Object type	RA, Dec	Approx. transit date at local midnight	Distance
Orion	Emission nebula	05h 41.7m, –01° 51′	December 26	1 thousand light years
Age	**Apparent size**	**Magnitude**	**Sky Atlas 2000.0 chart**	**Herald–Bobroff chart**
	30′		11	C53/D24

Nicknamed the "Flame Nebula" or the "Tank Track Nebula". This is a star-forming region. Professional telescopes show an embedded young open cluster. The number of binary stars in low-density star-forming regions is usually considerably higher than in the Galactic field (where probably > 50% of main-sequence stars are binaries). Although NGC 2024 is a dense star-forming region, it too has a large number of binary stars, with approximately 75% of the stars embedded in NGC 2024 being multiple stars.

NGC 2068 (M 78)

Constellation	Object type	RA, Dec	Approx. transit date at local midnight	Distance
Orion	Reflection nebula	05h 46.8m, +00° 05′	November 11	1.5 thousand light years
Age	**Apparent size**	**Magnitude**	**Sky Atlas 2000.0 chart**	**Herald–Bobroff chart**
	8′	8	11	C53/D24

M 78 is a star-forming region. It is illuminated by the double star HD 290862, which is the more southern of the two bright stars seen in this nebula in amateur telescopes. M 78 is part of a giant molecular cloud (the Orion B cloud) that has a size of about 8° in professional telescopic studies (and also includes the fellow star-forming regions NGC 2071, NGC 2023 and NGC 2024) and is part of the much larger Orion–Monoceros complex (see M 42). Its size of 8′ corresponds to a diameter of a few light years.

NGC 2099 (M 37)

Constellation	Object type	RA, Dec	Approx. transit date at local midnight	Distance
Auriga	Open cluster	05h 52.3m, +32° 33′	December 29	4 thousand light years
Age	**Apparent size**	**Magnitude**	**Sky Atlas 2000.0 chart**	**Herald–Bobroff chart**
4 hundred million years	24′	5.6	5	C35/D25

It has a diameter of about 30 light years. It lies close to M 38 and M 36 (see M 36). This is a rich cluster, with well over 2 thousand stars having been counted as members in professional telescopic studies.

NGC 2126

Constellation	Object type	RA, Dec	Approx. transit date at local midnight	Distance
Auriga	Open cluster	06h 02.5m, +49° 52′	December 31	5 thousand light years
Age	**Apparent size**	**Magnitude**	**Sky Atlas 2000.0 chart**	**Herald–Bobroff chart**
	6′	10.2	5	C17

It lies relatively far above the Galactic plane for an open cluster (1 thousand light years) at a distance of about 32 thousand light years from the Galactic center. It has a diameter of nearly 10 light years.

NGC 2129

Constellation	Object type	RA, Dec	Approx. transit date at local midnight	Distance
Gemini	Open cluster	06h 00.7m, +23° 19′	December 31	5 thousand light years
Age	**Apparent size**	**Magnitude**	**Sky Atlas 2000.0 chart**	**Herald–Bobroff chart**
20 million years	7′	6.7	5	C34

Recent work suggests that this may not actually be an open cluster, since many of the stars in this "cluster" are at a wide variety of distances. If proven true, this highlights the well-known hazards associated with telling the difference between stars that belong to a cluster from those that are simply field stars that happen to lie along the same line-of-sight but are at a different distance (see M 39 for further discussion).

NGC 2158

Constellation	Object type	RA, Dec	Approx. transit date at local midnight	Distance
Gemini	Open cluster	06h 07.4m, +24° 06′	January 1	16 thousand light years
Age	**Apparent size**	**Magnitude**	**Sky Atlas 2000.0 chart**	**Herald–Bobroff chart**
1 billion years	5′	8.6	5	C34

This is a populous cluster with almost 1 thousand member stars in professional telescopic studies. It lies almost directly in the Galactic anticenter direction (i.e. directly outward from us in the opposite direction from the center of the Galaxy). It is over 40 thousand light years from the center of the Galaxy, but lies only 4 hundred light years above the central plane of our Galaxy. It has a diameter of a little over 20 light years. M 35 (NGC 2168) lies only 24′ NE, but is not near NGC 2158 in space (the two lie roughly 13 thousand light years apart).

NGC 2168 (M 35)

Constellation	Object type	RA, Dec	Approx. transit date at local midnight	Distance
Gemini	Open cluster	06h 09.0m, +24° 21'	January 2	3 thousand light years
Age	**Apparent size**	**Magnitude**	**Sky Atlas 2000.0 chart**	**Herald–Bobroff chart**
2 hundred million years	28'	5.1	5	C34

It lies almost directly in the Galactic anticenter direction (i.e. directly outward from us in the opposite direction from the center of the Galaxy), about one hundred light years above the Galactic central plane. The open cluster NGC 2158 (see NGC 2158) lies only 24' SW, but is not near M 35 in space (NGC 2158 is roughly 13 thousand light years farther away). The total mass of M 35 is several thousand Suns.

NGC 2169

Constellation	Object type	RA, Dec	Approx. transit date at local midnight	Distance
Orion	Open cluster	06h 08.6m, +13° 58'	January 2	3 thousand light years
Age	**Apparent size**	**Magnitude**	**Sky Atlas 2000.0 chart**	**Herald–Bobroff chart**
50 million years	7'	5.9	11	C34

It contains the mag. 10.8 star GSC 00742–02169 near the center of the cluster (see Figure 2169). Professional telescopic studies show that this star is a variable, chemically peculiar "Ap" star (sometimes instead designated "CP2"), the variations (of < 0.1 magnitude) being caused by misalignment between the magnetic field poles and the rotation axis (making this a so-called "oblique rotator"), which is a common phenomenon with Ap stars. Ap stars have abnormally strong absorption in certain spectral lines (in this case due to the presence of silicon) – see NGC 7243. A more easily seen Ap star of this type is θ Aurigae (mag. 2.7), which is 23° N of this cluster.

NGC 2185

Constellation	Object type	RA, Dec	Approx. transit date at local midnight	Distance
Monoceros	Reflection nebula	06h 11.0m, –06° 14'	January 2	3 thousand light years
Age	**Apparent size**	**Magnitude**	**Sky Atlas 2000.0 chart**	**Herald–Bobroff chart**
	2'		11	C52

Three faint reflection nebulae are lined up nearby in a westerly direction: NGC 2183 (4.5' W), the brighter NGC 2182 (25' W), and NGC 2170 (55' W), which along with NGC 2185, all belong to the so-called Mon R2 association. This is a collection of reflection nebulae and B stars that are all within a few hundred light years of each other and are thought to be midway in the evolution process between a cold dust/gas cloud and an emission nebula/O star-forming region like the Orion Nebula (M 42).

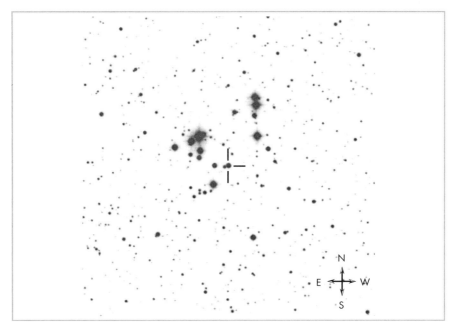

Figure 2169 The cross-hairs indicate the chemically peculiar (Ap or CP2) star GSC 00742-02169 in the open cluster NGC 2169. The area shown is 15' × 15'. From the Digitized Sky Survey (Space Telescope Science Institute) based on photographic data of the National Geographic Society – Palomar Observatory Sky Survey (POSS I).

NGC 2186

Constellation	Object type	RA, Dec	Approx. transit date at local midnight	Distance
Orion	Open cluster	06h 12.1m, +05° 28'	January 3	5 thousand light years
Age	**Apparent size**	**Magnitude**	**Sky Atlas 2000.0 chart**	**Herald–Bobroff chart**
50 million years	4'	8.7	11	C52/D23

It has a diameter of about 6 light years. It lies about 6 hundred light years below the Galactic central plane and was discovered in 1786 by William Herschel.

NGC 2194

Constellation	Object type	RA, Dec	Approx. transit date at local midnight	Distance
Orion	Open cluster	06h 13.8m, +12° 48'	January 3	12 thousand light years
Age	**Apparent size**	**Magnitude**	**Sky Atlas 2000.0 chart**	**Herald–Bobroff chart**
550 million years	10'	8.5	11	C52/D23

It has a diameter of about 30 light years and was discovered by William Herschel in 1784. About 80 member stars are counted in professional telescopic studies.

NGC 2204

Constellation	Object type	RA, Dec	Approx. transit date at local midnight	Distance
Canis Major	Open cluster	06h 15.5m, −18° 40′	January 3	9 thousand light years
Age	**Apparent size**	**Magnitude**	**Sky Atlas 2000.0 chart**	**Herald–Bobroff chart**
1 billion years	13′	8.6	11	C70

This is one of the furthest open clusters from the Galactic central plane (lying 3–4 thousand light years below it). It is over 30 light years in diameter. NGC 2211 (barred lenticular galaxy, 1.4′ × 0.7′, mag. 12.7) lies 39′ E, and forms a galaxy pair with the faint galaxy NGC 2212 (barred lenticular galaxy, 1.5′ × 0.8′, mag. 13.4, 2′ NE of NGC 2211). At 90 million light years distance, these two galaxies lie approximately 10 thousand times farther away from us than NGC 2204.

NGC 2215

Constellation	Object type	RA, Dec	Approx. transit date at local midnight	Distance
Monoceros	Open cluster	06h 20.8m, −07° 17′	January 5	4 thousand light years
Age	**Apparent size**	**Magnitude**	**Sky Atlas 2000.0 chart**	**Herald–Bobroff chart**
2 hundred million years	11′	8.4	11	C52

It has a diameter of a little over 10 light years and was discovered in 1785 by William Herschel.

NGC 2232

Constellation	Object type	RA, Dec	Approx. transit date at local midnight	Distance
Monoceros	Open cluster	06h 28.0m, −04° 51′	January 6	1 thousand light years
Age	**Apparent size**	**Magnitude**	**Sky Atlas 2000.0 chart**	**Herald–Bobroff chart**
50 million years	30′	3.9	11	C52

This is one of the nearest open clusters to us (only about 10 of the 1.5 thousand or so open clusters in our Galaxy are closer). It has a diameter of about 10 light years.

NGC 2244 (NGC 2237, 2238, 2239, 2244, 2246)

Constellation	Object type	RA, Dec	Approx. transit date at local midnight	Distance
Monoceros	Open cluster + emission nebula	06h 31.9m, +04° 57′	January 7	5 thousand light years

Age	Apparent size	Magnitude	Sky Atlas 2000.0 chart	Herald–Bobroff chart
2 million years	24′	4.8	11	C52/D23

The gas from which this young cluster formed is visible as the surrounding emission nebula, called the "Rosette Nebula", which extends out to a diameter of 60–80′. Bits and pieces of the Rosette Nebula complex have been given individual NGC numbers and include NGC 2237, 2238, 2239, 2244 and 2246. The entire Rosette Nebula/cluster has a mass of about 330 thousand Suns, of which about 11 thousand solar masses is dust (the rest is essentially gas). The nebula is thought to be roughly a spherical shell with a diameter of several hundred light years, with the "donut hole" shape we see actually being a hole in the near side of this shell. The cluster contains about 130 stars in professional telescopic studies.

NGC 2251

Constellation	Object type	RA, Dec	Approx. transit date at local midnight	Distance
Monoceros	Open cluster	06h 34.6m, +08° 22′	January 8	5 thousand light years

Age	Apparent size	Magnitude	Sky Atlas 2000.0 chart	Herald–Bobroff chart
3 hundred million years	10′	7.3	12	C52/D23

It lies very close to the Galactic central plane (10 light years above it). It has a diameter of about 15 light years.

NGC 2261

Constellation	Object type	RA, Dec	Approx. transit date at local midnight	Distance
Monoceros	Reflection nebula	06h 39.2m, +08° 45′	January 9	2.5 thousand light years

Age	Apparent size	Magnitude	Sky Atlas 2000.0 chart	Herald–Bobroff chart
	2.0′ × 1.0′		12	C52/D23

Nicknamed "Hubble's Variable Nebula". This fan-shaped reflection nebula is illuminated by the variable star R Mon (which lies at the south end of the nebula but is
obscured at optical wavelengths by the nebula itself). R Mon is a Be star (see M 47/NGC
2422 for explanation) and professional telescopes find it is a binary star (with separation 0.7″). The nebula that R Mon is lighting up is thought to consist of a bipolar shell,
with the southern half not visible in amateur telescopes. The northern half of the shell
that we see as NGC 2261 resembles a parabola that has been rotated about a nearly
north–south line. In other words, it is like a wine-glass without its stem, with R Mon
at the bottom of the wine-glass. The parabolic shell is thought to be thin (0.01 light
years thick, which is 8 hundred times the Earth–Sun distance) and nearly edge-on with
its symmetry axis inclined from our line-of-sight by about 70°. In other words, we are
looking at its wine-glass shape nearly side-on with its bottom on the south end. The
parabolic shell may be the result of material being ejected from an accretion disk that
surrounds R Mon. Filaments of this material are thought to cast shadows on the walls
of the shell, resulting in the variable appearance of the nebula that gives it its nickname.
R Mon and the nebula it has created are thought to be only a few hundred thousand
years old. They may be an outlier of the open cluster NGC 2264 (see NGC 2264), but
this is not certain.

NGC 2264

Constellation	Object type	RA, Dec	Approx. transit date at local midnight	Distance
Monoceros	Open cluster + emission nebula	06h 41.0m, +09° 54′	January 10	2.5 thousand light years

Age	Apparent size	Magnitude	Sky Atlas 2000.0 chart	Herald–Bobroff chart
5 million years	20′	3.9	12	C52/D23

It is nicknamed the "Christmas Tree Cluster" due to its apparent shape when viewed
with south at the top of the eyepiece. This is a star-forming region containing about 4
hundred member stars in professional telescopic studies. Although the nebulosity in
which the cluster is embedded is largely ionized hydrogen, directly behind this (and not
optically visible) is a cloud of molecules (including ammonia, carbon monoxide,
formaldehyde) having a mass of 10 thousand Suns. At the south end of the cluster (i.e.
just south of the tip of the Christmas tree) is the "Cone Nebula". This is a cone-shaped
dark molecular cloud (5′ × 10′) and a challenging object in large amateur telescopes. It
may be the remains of what was once a larger cloud that has been disintegrated by
nearby star-forming activity, similar to the "Pillars of Creation" region in M 16 (see M
16/NGC 6611). The cone-shape appears to have been protected from stellar winds by a
dense globule at the tip of the cone. This cluster is so young that about half of its stars
still have circumstellar disks (see NGC 2362).

NGC 2266

Constellation	Object type	RA, Dec	Approx. transit date at local midnight	Distance
Gemini	Open cluster	06h 43.3m, +26° 58′	January 10	11 thousand light years
Age	**Apparent size**	**Magnitude**	**Sky Atlas 2000.0 chart**	**Herald–Bobroff chart**
6 hundred million years	7′	9.5	5	C34

It contains at least 190 member stars in professional telescopic studies. It is by far the youngest cluster that sits so far from the Galactic central plane (it lies 2 thousand light years above this plane, with other clusters this far off the disk being at least twice as old). It has a diameter of about 20 light years.

NGC 2281

Constellation	Object type	RA, Dec	Approx. transit date at local midnight	Distance
Auriga	Open cluster	06h 48.3m, +41° 05′	January 12	2 thousand light years
Age	**Apparent size**	**Magnitude**	**Sky Atlas 2000.0 chart**	**Herald–Bobroff chart**
9 hundred million years	15′	5.4	5	C34

It contains over one hundred member stars in professional telescopic studies and has a diameter of a little under 10 light years.

NGC 2286

Constellation	Object type	RA, Dec	Approx. transit date at local midnight	Distance
Monoceros	Open cluster	06h 47.7m, –03° 09′	January 12	4 thousand light years
Age	**Apparent size**	**Magnitude**	**Sky Atlas 2000.0 chart**	**Herald–Bobroff chart**
8 hundred million years	15′	7.5	12	C52

It has a diameter of about 15 light years and was discovered by William Herschel in 1785. Professional telescopic studies count about one hundred member stars.

NGC 2287 (M 41)

Constellation	Object type	RA, Dec	Approx. transit date at local midnight	Distance
Canis Major	Open cluster	06h 46.0m, –20° 45′	January 11	2 thousand light years
Age	**Apparent size**	**Magnitude**	**Sky Atlas 2000.0 chart**	**Herald–Bobroff chart**
2 hundred million years	38′	4.5	19	C70/D22

It has a diameter of about 20 light years. A large percentage (perhaps as high as 80%) of its stars are binary stars. Open clusters are beasts of the Galactic disk, and are stripped of their stars over time by gravitational interaction with material in the disk as they jostle about the disk while rotating with it. For an open cluster in our vicinity of the Galaxy, like this one, typical lifetimes are thought to be a little over 1/2 billion years, so that this cluster is approaching middle age. The cluster has about 70 members with magnitude brighter than 12, although our view of the cluster to this magnitude is "contaminated" by about as many field stars as cluster members (see M 39).

NGC 2301

Constellation	Object type	RA, Dec	Approx. transit date at local midnight	Distance
Monoceros	Open cluster	06h 51.8m, +00° 28′	January 13	3 thousand light years
Age	**Apparent size**	**Magnitude**	**Sky Atlas 2000.0 chart**	**Herald–Bobroff chart**
2 hundred million years	12′	6.0	12	C52/D23

The cluster region has many field stars (i.e. background and foreground stars that are not members of the cluster but which happen to lie along the same line-of-sight as the cluster – see M 39/NGC 7092 for further discussion of this). Indeed, of 9 hundred stars studied in professional telescopic studies to mag. 17 within the cluster region, only about one hundred are considered members. The cluster is about 10 light years in diameter.

NGC 2304

Constellation	Object type	RA, Dec	Approx. transit date at local midnight	Distance
Gemini	Open cluster	06h 55.2m, +18° 00′	January 13	13 thousand light years
Age	**Apparent size**	**Magnitude**	**Sky Atlas 2000.0 chart**	**Herald–Bobroff chart**
8 hundred million years	5′	10.0	12	C34

It has a diameter of about 20 light years and was discovered by William Herschel in 1783. It is located in the direction of the Galactic anticenter. The region between us and this cluster has low levels of interstellar light extinction, since intervening dust dims the stars in this cluster by only 0.08 magnitudes per kiloparsec (3.26 thousand light years). This is much less than the average two magnitudes of dimming for every kiloparsec that is typical when light travels in the central plane of our Galaxy.

NGC 2311

Constellation	Object type	RA, Dec	Approx. transit date at local midnight	Distance
Monoceros	Open cluster	06h 57.8m, −04° 37′	January 14	7 thousand light years
Age	**Apparent size**	**Magnitude**	**Sky Atlas 2000.0 chart**	**Herald–Bobroff chart**
4 hundred million years	7′	9.6	12	C52

It has a diameter of about 15 light years and was discovered by William Herschel in 1786.

NGC 2323 (M 50)

Constellation	Object type	RA, Dec	Approx. transit date at local midnight	Distance
Monoceros	Open cluster	07h 02.1m, –08° 23′	January 15	3 thousand light years
Age	**Apparent size**	**Magnitude**	**Sky Atlas 2000.0 chart**	**Herald–Bobroff chart**
1 hundred million years	16′	5.9	12	C52/D22

It has roughly 2 hundred members, a diameter of about 15 light years and lies about 70 light years below the central Galactic plane. The interstellar gas and dust in the disk of our Galaxy is not uniformly distributed at light year length scales, but instead is clumped into patches with typical masses of a few hundred Suns. Open clusters are thought to form from such interstellar clouds of gas and dust.

NGC 2324

Constellation	Object type	RA, Dec	Approx. transit date at local midnight	Distance
Monoceros	Open cluster	07h 04.1m, +01° 03′	January 16	13 thousand light years
Age	**Apparent size**	**Magnitude**	**Sky Atlas 2000.0 chart**	**Herald–Bobroff chart**
6 hundred million years	8′	8.4	12	C52

It has a diameter of about 30 light years, and contains over 130 stars according to professional telescopic studies.

NGC 2335

Constellation	Object type	RA, Dec	Approx. transit date at local midnight	Distance
Monoceros	Open cluster	07h 06.8m, –10° 02′	January 16	3 thousand light years
Age	**Apparent size**	**Magnitude**	**Sky Atlas 2000.0 chart**	**Herald–Bobroff chart**
2 hundred million years	12′	7.2	12	C52/D22

It has a diameter of about 10 light years. It has been proposed that this cluster and NGC 2343 (42′ SE – see NGC 2343) form a double system. The emission + reflection nebula Gum 1 (20′ in diameter) lies 38′ E and 26′S.

NGC 2343

Constellation	Object type	RA, Dec	Approx. transit date at local midnight	Distance
Monoceros	Open cluster	07h 08.1m, –10° 37′	January 17	3 thousand light years
Age	**Apparent size**	**Magnitude**	**Sky Atlas 2000.0 chart**	**Herald–Bobroff chart**
1 hundred million years	7′	6.7	12	C52/D22

It has a diameter of about 7 light years. The faint emission nebula IC 2177 (120′ × 40′) is centered 45′ to the east.

NGC 2353

Constellation	Object type	RA, Dec	Approx. transit date at local midnight	Distance
Monoceros	Open cluster	07h 14.5m, −10° 16′	January 18	4 thousand light years

Age	Apparent size	Magnitude	Sky Atlas 2000.0 chart	Herald–Bobroff chart
1 hundred million years	20′	7.1	12	C52/D22

It has a diameter of about 20 light years. It contains over 2 hundred stars to mag. 21 in professional telescopic studies. It lies only 20 or so light years above the Galactic central plane at a distance of about 30 thousand light years from the center of our Galaxy. Even though it lies at about the same distance as the cluster, the brightest star in the field (HD 55879, mag. 6.0) is not thought to be a member of this cluster. HD 55879 is instead a much younger star that is part of a group of young O and B stars called the Canis Major OB1 stellar association that is spread out over a 4° × 4° area centered near this location.

NGC 2354

Constellation	Object type	RA, Dec	Approx. transit date at local midnight	Distance
Canis Major	Open cluster	07h 14.3m, −25° 42′	January 18	6 thousand light years

Age	Apparent size	Magnitude	Sky Atlas 2000.0 chart	Herald–Bobroff chart
1 hundred million years	20′	6.5	19	C70/D21

It has a diameter somewhat over 30 light years. A magnitude 11.8 eclipsing binary variable star (QU CMa, or GSC 06528–01240) in the SW part of this cluster (see Figure 2354) is also a blue straggler (see NGC 6633 for the meaning of "blue straggler"). It is one of only a few confirmed binary blue stragglers. Its "blueness" may be due to mass exchange between stars in this close binary pair, with mass accreting from the small, red, dim secondary star onto the brighter (blue straggler) primary star.

NGC 2355

Constellation	Object type	RA, Dec	Approx. transit date at local midnight	Distance
Gemini	Open cluster	07h 17.0m, +13° 45′	January 19	5 thousand light years

Age	Apparent size	Magnitude	Sky Atlas 2000.0 chart	Herald–Bobroff chart
1 billion years	9′	9.7	12	C51

As with all relatively old open clusters, NGC 2355 stays fairly far out in the Galaxy, never approaching closer than about 30 thousand light years to the Galactic center. Its secret to long life is its avoidance of other Galactic matter that is more concentrated at locations both closer in to the Galactic center and in the Galactic disk. Indeed, in its approximately four or so revolutions around the Galaxy in its lifetime, it has spent less than 10% or so of its time in the thin layer of the Galactic disk (150 light years in thickness) where encounters with large molecular clouds could begin to disrupt it, having crossed this disk about 40 times. In its present location it is at its maximum distance above the Galactic disk (1.1 thousand light years) and is about to reverse direction to head below the Galactic disk.

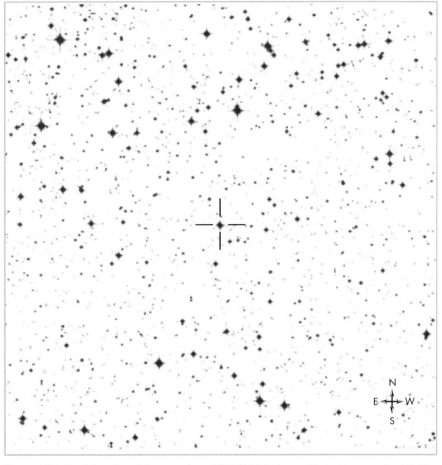

Figure 2354 The SW portion of the open cluster NGC 2354 is shown with the blue straggler QU CMa marked by cross-hairs. The area shown is 15′ × 15′. From the Digitized Sky Survey (Space Telescope Science Institute) based on photographic data of the National Geographic Society – Palomar Observatory Sky Survey (POSS I).

NGC 2359

Constellation	Object type	RA, Dec	Approx. transit date at local midnight	Distance
Canis Major	Emission nebula	07h 18.5m, −13° 14′	January 19	16 thousand light years
Age	Apparent size	Magnitude	Sky Atlas 2000.0 chart	Herald–Bobroff chart
	10′ × 5′		12	C52/D22

Its appearance has resulted in the nickname "Thor's Helmet" or the "Duck Nebula". It is an ionized hydrogen (HII) region surrounding a WN Wolf–Rayet star (designated HD 56925 or HIP 35378 and lying at mag. 11.4 near the middle of the northern ring-like segment of the nebula – see Figure 2359 (overleaf); see NGC 2403 for more on Wolf–Rayet stars). The visible nebula is part of a larger neutral gas cloud, the HII region fluorescing due to ionizing radiation from the Wolf–Rayet star. The stellar wind from this star has blown out a "bubble" in the HII region that gives a 4.5′ ring-like appearance to the northern segment in professional telescopic studies. The ionized (visible) nebula is thought to have a mass of a few hundred Suns, with several times this amount of mass contained in the adjacent (nonvisible) neutral gas clouds (with most of the neutral gas lying at the south end of the nebula, south of the duck's "bill").

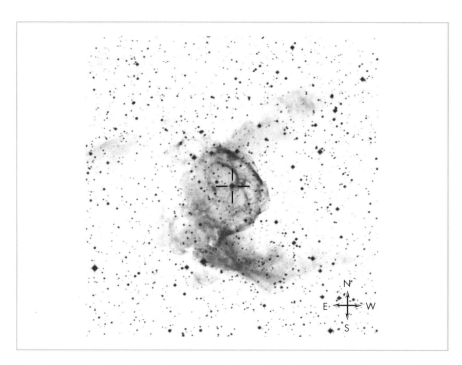

Figure 2359 The emission nebula NGC 2359 is shown with the ionizing Wolf–Rayet star marked by cross-hairs. The area shown is 15′ × 15′. From the Digitized Sky Survey (Space Telescope Science Institute) based on photographic data of the National Geographic Society – Palomar Observatory Sky Survey (POSS I).

NGC 2360

Constellation	Object type	RA, Dec	Approx. transit date at local midnight	Distance
Canis Major	Open cluster	07h 17.7m, −15° 38′	January 19	4 thousand light years
Age	**Apparent size**	**Magnitude**	**Sky Atlas 2000.0 chart**	**Herald–Bobroff chart**
1 billion years	13′	7.2	12	C52/D22

It has a diameter of about 15 light years and was discovered in 1783 by Caroline Herschel (William Herschel's sister).

NGC 2362

Constellation	Object type	RA, Dec	Approx. transit date at local midnight	Distance
Canis Major	Open cluster	07h 18.7m, −24° 57′	January 19	4 thousand light years

Age	Apparent size	Magnitude	Sky Atlas 2000.0 chart	Herald–Bobroff chart
5 million years	8′	4.1	19	C70/D21

About 12% of the stars in this young cluster still have circumstellar disks around them that formed as these stars collapsed from the surrounding gas and dust. Since this cluster is approximately 5 million years old and started with a much higher fraction of stars with disks, it is thus thought that planets must form within about 5 million years of the parent star's formation, since planets form from circumstellar disks and make these disks disappear. The brightest star in this cluster (τ CMa, mag. 4.1) is actually a quadruple star system in professional telescopes, with three of the stars in this system having masses around 50 times that of our Sun and radii nearly 10 times that of our Sun. This quadruple system is thought to have perhaps formed when two binary systems merged.

NGC 2371

Constellation	Object type	RA, Dec	Approx. transit date at local midnight	Distance
Gemini	Planetary nebula	07h 25.6m, +29° 29′	January 21	3 thousand light years

Age	Apparent size	Magnitude	Sky Atlas 2000.0 chart	Herald–Bobroff chart
	55″	12	5	C33

It has a diameter of about 1 light year. The central star (mag. 14.8) is a Wolf–Rayet star (see NGC 40 for explanation) and is visible in large amateur telescopes between the two lobes of this bipolar planetary nebula (see M 27/NGC 6853 for explanation of "bipolar" planetary nebulae). The two lobes have been given individual NGC numbers (NGC 2371 and 2372), but are parts of the same planetary nebula (see NGC 2372).

NGC 2372

Constellation	Object type	RA, Dec	Approx. transit date at local midnight	Distance
Gemini	Planetary nebula	07h 25.6m, +29° 30′	January 21	3 thousand light years

Age	Apparent size	Magnitude	Sky Atlas 2000.0 chart	Herald–Bobroff chart
	55″	12	5	C33

It has a diameter of about 1 light year. This is the other, dimmer (NE) half of the same planetary nebula of which NGC 2371 is the SW half (see NGC 2371).

NGC 2392

Constellation	Object type	RA, Dec	Approx. transit date at local midnight	Distance
Gemini	Planetary nebula	07h 29.2m, +20° 55′	January 22	5 thousand light years

Age	Apparent size	Magnitude	Sky Atlas 2000.0 chart	Herald– Bobroff chart
	16″	9	5	C34

Nicknamed the "Clown Face Nebula" or the "Eskimo Nebula". Professional telescopic studies suggest that this nebula consists of a 16″ diameter bright, inner, prolate spheroid (i.e. cigar shape – see NGC 2022) viewed nearly pole-on and expanding at about 60 km/s. This is surrounded by an outer shell that is thought to be an oblate spheroid (like a squashed tennis ball or a discus) expanding at < 20 km/s. Thus, this nebula consists of a cigar shaped region exploding into a discus shaped region, with the ends of the cigar sticking out the flat sides of the discus. The outer shell contains a fringe of filaments and knots that lie in a roughly planar disk region (and make up the "hood" of the "Eskimo" face). The fringe may be due to shocks where high-speed pulsed jets exit the inner region and interact with the outer shell. It is believed that in 6 thousand years or so the inner shell will have completely expanded through the outer shell, after which time this nebula may evolve into a bipolar planetary nebula (see M 27/NGC 6853). The central star (mag. 10.5), which shed the complex mess of gas making up the nebula, is visible in amateur telescopes.

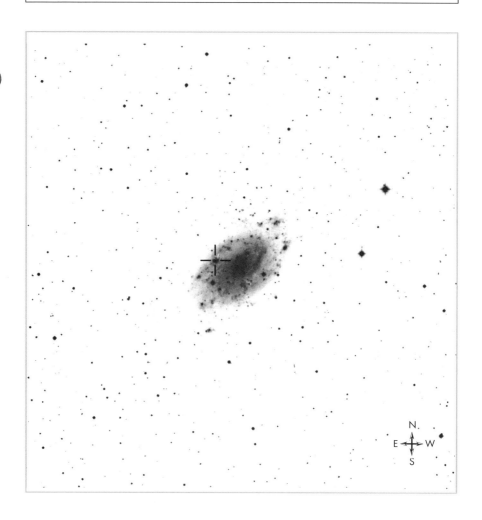

NGC 2395

Constellation	Object type	RA, Dec	Approx. transit date at local midnight	Distance
Gemini	Open cluster	07h 27.2m, +13° 37′	January 22	4 thousand light years
Age	**Apparent size**	**Magnitude**	**Sky Atlas 2000.0 chart**	**Herald–Bobroff chart**
1 billion years	12′	8.0	12	C33

It has a diameter of about 15 light years. It lies 1 thousand light years above the Galactic central plane, which is relatively far for an open cluster.

NGC 2403

Constellation	Object type	RA, Dec	Approx. transit date at local midnight	Distance
Camelopardalis	Barred spiral galaxy	07h 36.8m, +65° 36′	January 24	10 million light years
Age	**Apparent size**	**Magnitude**	**Sky Atlas 2000.0 chart**	**Herald–Bobroff chart**
	23.4′ × 11.8′	8.5	1	C15

It has an optical diameter of about 70 thousand light years. It is thought to be part of the M 81 group of gravitationally bound galaxies, which includes galaxies in a region with a radius of about 5 million light years (see M 81/NGC 3031). The galaxy rotates (at speeds up to 170 km/s at a distance of 10″ from its center) about an axis that is inclined from our line-of-sight at about 60°. It has a morphology similar to M 33 (see M 33/NGC 598). It contains several compact star-forming regions that are bright ionized hydrogen emission (HII) regions. The brightest of these is 2′ E of the center of this galaxy and has its own NGC number (NGC 2404 – see Figure 2403); this HII region is 2 thousand light years in diameter and contains 30–40 Wolf–Rayet stars (both WN and WC types – see NGC 40 for explanation). These "Population I" Wolf–Rayet stars in NGC 2403 are far more massive (with masses several tens of times that of our Sun) than the Wolf–Rayet stars that occur in planetary nebulae like NGC 40 (which have a mass less than our Sun). Population I Wolf–Rayets are one of the evolutionary stages undergone by massive stars (which live only a few million years before blowing up in supernovae, so that this HII region must be only a few million years old).

◄ **Figure 2403** (facing page) The cross-hairs mark the brightest HII region (NGC 2404) in the galaxy NGC 2403. The area shown is 30′ × 30′. From the Digitized Sky Survey (Space Telescope Science Institute) based on photographic data of the National Geographic Society – Palomar Observatory Sky Survey (POSS I).

NGC 2419

Constellation	Object type	RA, Dec	Approx. transit date at local midnight	Distance
Lynx	Globular cluster	07h 38.1m, +38° 53′	January 24	3 hundred thousand light years
Age	**Apparent size**	**Magnitude**	**Sky Atlas 2000.0 chart**	**Herald–Bobroff chart**
12–14 billion years	6.2′	10.4	5	C33

Once thought to be an intergalactic cluster (thus its nickname "The Intergalactic Wanderer"), it is now recognized as being gravitationally bound to our Galaxy, although it lies far out in the halo of our Galaxy (see NGC 7006 for explanation of "halo"). It is thought to have formed at the same time as the earliest star formation in our Galaxy from a supergiant molecular cloud that had an initial mass between about one hundred million and one billion solar masses. This cloud was part of the protogalaxy from which our Galaxy formed. NGC 2419 is one of the five most luminous globular clusters in our Galaxy. It has a mass of well over one million Suns.

NGC 2420

Constellation	Object type	RA, Dec	Approx. transit date at local midnight	Distance
Gemini	Open cluster	07h 38.4m, +21° 34′	January 24	9 thousand light years
Age	**Apparent size**	**Magnitude**	**Sky Atlas 2000.0 chart**	**Herald–Bobroff chart**
2 billion years	10′	8.3	5	C33

It contains nearly 7 hundred member stars in professional telescopic studies, with most of the stars having masses near that of our Sun. It has a diameter of about 25 light years (but member stars extend out to 4 times this distance in professional telescopes). This cluster orbits the Galactic center at a radial distance that has varied from about 28 to 38 thousand light years during its lifetime (it presently sits about 35 thousand light years from the Galactic center), although it spends more of its time near the farther out of these two distances. Its orbit always stays within about 3 thousand light years above and below the Galactic central plane (it currently sits near its maximum distance above the Galactic central plane). Like all older open clusters that spend their lives fairly far from the Galactic center, it contains lower amounts of compounds heavier than helium (i.e. "metals"). In fact, the farther out an open cluster is from the Galactic center, the less "metal" its stars contain. The "metallicity" of NGC 2420 is about 1/3 that of our Sun.

NGC 2421

Constellation	Object type	RA, Dec	Approx. transit date at local midnight	Distance
Puppis	Open cluster	07h 36.2m, −20° 37′	January 24	7 thousand light years
Age	**Apparent size**	**Magnitude**	**Sky Atlas 2000.0 chart**	**Herald–Bobroff chart**
20 million years	10′	8.3	19	C69/D22

It has a diameter of about 20 light years and was discovered in 1799 by William Herschel.

NGC 2422 (M 47)

Constellation	Object type	RA, Dec	Approx. transit date at local midnight	Distance
Puppis	Open cluster	07h 36.6m, −14° 29′	January 24	2 thousand light years
Age	**Apparent size**	**Magnitude**	**Sky Atlas 2000.0 chart**	**Herald–Bobroff chart**
1 hundred million years	30′	4.4	12	C51/D22

Its diameter is about 18 light years. Interstellar dust between us and the stars in this cluster cause its stars to appear dimmer by only a few tenths of a magnitude, which is much less than the average two magnitudes of dimming for every kiloparsec (3.26 thousand light years) that is typical when light travels in the central plane of our Galaxy. Several Be stars are known in this cluster, the brightest of which is HD 60856 at mag. 8 and is readily visible in amateur telescopes (see Figure 2422). Be stars are B-type stars that are peculiar because of hydrogen Balmer emission lines in their spectra, due to atomic transitions in material expelled by high rotational velocities into a circumstellar disk in the equatorial plane of the star.

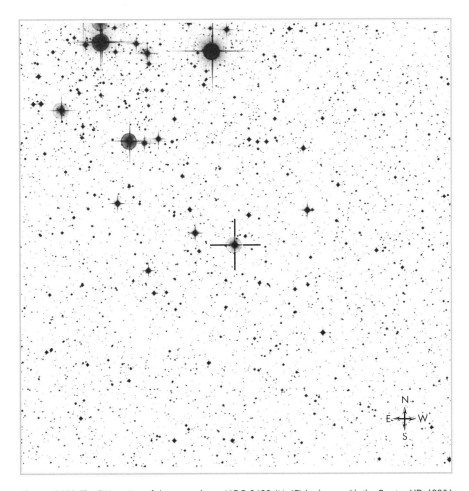

Figure 2422 The SW portion of the open cluster NGC 2422 (M 47) is shown with the Be star HD 60856 in the center of the figure (marked by cross-hairs, not diffraction spikes like the other stars not in the center of the figure). The area shown is 30′ × 30′. From the Digitized Sky Survey (Space Telescope Science Institute) based on photographic data of the National Geographic Society – Palomar Observatory Sky Survey (POSS I).

NGC 2423

Constellation	Object type	RA, Dec	Approx. transit date at local midnight	Distance
Puppis	Open cluster	07h 37.1m, −13° 52'	January 24	2 thousand light years
Age	**Apparent size**	**Magnitude**	**Sky Atlas 2000.0 chart**	**Herald–Bobroff chart**
5 hundred million years	19'	6.7	12	C51/D22

This cluster contains 71 members down to mag. 13 in professional photometric studies. It has a diameter of about 10 light years. It lies a little over one hundred light years above the Galactic central plane. M 47 (NGC 2422) is 39' S but the two are about 0.5–1 thousand light years apart.

NGC 2437 (M 46)

Constellation	Object type	RA, Dec	Approx. transit date at local midnight	Distance
Puppis	Open cluster	07h 41.8m, −14° 49'	January 25	5 thousand light years
Age	**Apparent size**	**Magnitude**	**Sky Atlas 2000.0 chart**	**Herald–Bobroff chart**
3 hundred million years	27'	6.1	12	C51/D22

Its 27' size corresponds to a diameter of about 40 light years. It lies about 4 hundred light years above the Galactic central plane at a distance of about 35 thousand light years from the Galactic center. Although the planetary nebula NGC 2438 lies on the NE edge of M 46 (see NGC 2438), their differing relative velocities and the young age of M 46 together suggest that NGC 2438 is not part of M 46. Whether NGC 2438 is a fore- ground or background object remains uncertain (it has been suggested it is a foreground object lying about 2 thousand light years closer than M 46).

NGC 2438

Constellation	Object type	RA, Dec	Approx. transit date at local midnight	Distance
Puppis	Planetary nebula	07h 41.8m, −14° 44'	January 25	3 thousand light years
Age	**Apparent size**	**Magnitude**	**Sky Atlas 2000.0 chart**	**Herald–Bobroff chart**
	66"	11.0	12	C51/D22

Its diameter of 1.1' implies an actual diameter of about 1 light year, but this nebula extends out to about 4 times this size in professional telescopic images in a spherical halo outside the inner spherical portion visible in amateur telescopes. The nebula is thought to have begun its formation about 45 thousand years ago, and so is considered an old planetary nebula. Indeed, its halo is undergoing "recombination", a late stage in the life of a planetary nebula whereby atoms that were previously photoionized by radi- ation from the central star (which is not visible in amateur telescopes) are recombining with electrons into neutral atoms because the central star's luminosity has decreased dramatically. Although NGC 2438 lies on the NE edge of the open cluster M 46 (see M 46/NGC 2437), their differing relative velocities and the young age of M 46 together suggest that NGC 2438 is not part of M 46. Whether it is a foreground or background object remains uncertain (it has been suggested it is a foreground object lying about 2 thousand light years closer than M 46).

NGC 2440

Constellation	Object type	RA, Dec	Approx. transit date at local midnight	Distance
Puppis	Planetary nebula	07h 41.9m, –18° 13′	January 25	4 thousand light years

Age	Apparent size	Magnitude	Sky Atlas 2000.0 chart	Herald–Bobroff chart
	54″ × 20″	11	12	C69/D22

Professional telescopic studies find that this planetary nebula consists of two interlocking bipolar structures (i.e. two "hourglass" structures joined at the "waist" of the hourglass), one with position angle of 85° and inclined at a 40° angle from our line-of-sight, the other with position angle of 35° and lined up at a 90° angle from our line-of-sight. The two sets of bipolar lobes are thought to have been caused by two separate mass ejection events from the central star that ejected the lobes in different directions, perhaps because the circumstellar toroidal ring of material that leads to the formation of the jets (see M 27/NGC 6853) was oriented in different directions (possibly due to precession of this ring). It is one of only seven known planetary nebulae with a double bipolar (i.e. quadrapolar) structure. The nebula is thought to be 5–10 thousand years old.

NGC 2447 (M 93)

Constellation	Object type	RA, Dec	Approx. transit date at local midnight	Distance
Puppis	Open cluster	07h 44.5m, –23° 51′	January 26	3.5 thousand light years

Age	Apparent size	Magnitude	Sky Atlas 2000.0 chart	Herald–Bobroff chart
4 hundred million years	22′	6.2	19	C69/D21

It lies just above the Galactic central plane (by a few light years) a few thousand light years farther from the Galactic center than we are. It has a diameter of about 20 light years.

NGC 2479

Constellation	Object type	RA, Dec	Approx. transit date at local midnight	Distance
Puppis	Open cluster	07h 55.1m, –17° 42′	January 29	

Age	Apparent size	Magnitude	Sky Atlas 2000.0 chart	Herald–Bobroff chart
	7′	9.6	12	C69/D22

It was discovered in 1790 by William Herschel. Little research has been done on this cluster to date.

NGC 2482

Constellation	Object type	RA, Dec	Approx. transit date at local midnight	Distance
Puppis	Open cluster	07h 55.2m, −24° 15′	January 29	3 thousand light years

Age	Apparent size	Magnitude	Sky Atlas 2000.0 chart	Herald–Bobroff chart
4 hundred million years	12′	7.3	19	C69/D21

It has a diameter of about 10 light years. It lies less than one hundred light years above the Galactic central plane at a distance of about 29 thousand light years from the center of our Galaxy.

NGC 2489

Constellation	Object type	RA, Dec	Approx. transit date at local midnight	Distance
Puppis	Open cluster	07h 56.3m, −30° 04′	January 29	5 thousand light years

Age	Apparent size	Magnitude	Sky Atlas 2000.0 chart	Herald–Bobroff chart
2 hundred million years	5′	7.0	19	C69/D21

Its diameter is about 10 light years. It was discovered in 1785 by William Herschel.

NGC 2506

Constellation	Object type	RA, Dec	Approx. transit date at local midnight	Distance
Monoceros	Open cluster	08h 00.0m, −10° 46′	January 30	11 thousand light years

Age	Apparent size	Magnitude	Sky Atlas 2000.0 chart	Herald–Bobroff chart
2 billion years	7′	7.6	12	C51/D22

It contains over 8 hundred stars in professional telescopic studies. It has a diameter of about 20 light years. The initial mass of this cluster is believed to have been at least about 2 thousand solar masses. It has not strayed far from its birth location in the Galaxy, having stayed within a region about 2 thousand light years in diameter from this location, centered on the central Galactic plane at a radial distance of about 36 thousand light years from the Galactic center. This may explain why it remains so rich for its old age – it has avoided encounters with large gas clouds that tend to strip clusters of their members.

NGC 2509

Constellation	Object type	RA, Dec	Approx. transit date at local midnight	Distance
Puppis	Open cluster	08h 00.8m, −19° 03′	January 30	
Age	**Apparent size**	**Magnitude**	**Sky Atlas 2000.0 chart**	**Herald–Bobroff chart**
	8′	9.3	12	C69/D22

It was discovered in 1783 by William Herschel. Little research has been done on this cluster to date.

NGC 2527

Constellation	Object type	RA, Dec	Approx. transit date at local midnight	Distance
Puppis	Open cluster	08h 05.0m, −28° 09′	January 31	2 thousand light years
Age	**Apparent size**	**Magnitude**	**Sky Atlas 2000.0 chart**	**Herald–Bobroff chart**
5 hundred million years	22′	6.5	20	C69/D21

It has a diameter of about 10 light years. This is a relatively loose cluster, with the stars spaced an average of about 3 light years apart, although this is still considerably denser than the average 10 light year interstellar spacing in the neighborhood of our Sun.

NGC 2539

Constellation	Object type	RA, Dec	Approx. transit date at local midnight	Distance
Puppis	Open cluster	08h 10.6m, −12° 49′	February 2	4 thousand light years
Age	**Apparent size**	**Magnitude**	**Sky Atlas 2000.0 chart**	**Herald–Bobroff chart**
6 hundred million years	22′	6.5	12	C51

In professional telescopic studies it contains nearly 2 hundred stars, which are spaced an average of 5 light years apart. This interstellar spacing is relatively large for an open cluster, but still gives nearly ten times as many stars per unit volume as in our Galactic neighborhood. It has a diameter of about 30 light years.

NGC 2548 (M 48)

Constellation	Object type	RA, Dec	Approx. transit date at local midnight	Distance
Hydra	Open cluster	08h 13.7m, –05° 45′	February 2	2 thousand light years
Age	**Apparent size**	**Magnitude**	**Sky Atlas 2000.0 chart**	**Herald–Bobroff chart**
3 hundred million years	54′	5.8	12	C51

Professional telescopic studies find 165 stars with membership probability > 70%. It lies about 29 thousand light years from the center of our Galaxy and a little more than 5 hundred light years above the Galactic central plane. Its current position is close to its maximum excursion of 6 hundred light years from this plane, having crossed it eight times in its lifetime, while making slightly more than one revolution about the Galactic center. It has a diameter of about 30 light years. Messier's discovery and subsequent listing of this object resulted in an error in its quoted position, so that its location in some old star charts (before T.F. Morris' correction of this error in 1959) is incorrect.

NGC 2567

Constellation	Object type	RA, Dec	Approx. transit date at local midnight	Distance
Puppis	Open cluster	08h 18.5m, –30° 38′	February 4	5 thousand light years
Age	**Apparent size**	**Magnitude**	**Sky Atlas 2000.0 chart**	**Herald–Bobroff chart**
3 hundred million years	11′	7.0	20	C69/D21

It has a diameter of about 15 light years. The region between us and this cluster has low levels of interstellar light extinction, since the stars in this cluster are dimmed by only a few tenths of a magnitude by intervening dust. This is much less than the average two magnitudes of dimming for every kiloparsec (3.26 thousand light years) that is typical when light travels in the central plane of our Galaxy.

NGC 2571

Constellation	Object type	RA, Dec	Approx. transit date at local midnight	Distance
Puppis	Open cluster	08h 18.9m, –29° 45′	February 4	4 thousand light years
Age	**Apparent size**	**Magnitude**	**Sky Atlas 2000.0 chart**	**Herald–Bobroff chart**
50 million years	13′	7.0	20	C69/D21

It has a diameter of about 15 light years. It lies about 29 thousand light years from the center of our Galaxy and 250 light years above the central Galactic plane. The cluster does not have a high stellar density. Indeed, to magnitude 18 there are 34 stars per square arcminute in the inner 4′ of the cluster, which is only 17% more dense than the background stellar density.

NGC 2613

Constellation	Object type	RA, Dec	Approx. transit date at local midnight	Distance
Pyxis	Spiral galaxy	08h 33.4m, –22° 58′	February 7	85 million light years
Age	**Apparent size**	**Magnitude**	**Sky Atlas 2000.0 chart**	**Herald–Bobroff chart**
	7.1′ × 1.6′	10.3	20	C69

It rotates about an axis that is inclined at about 80° from our line-of-sight (with the south side closer to us). This is a massive galaxy, having a dynamical mass of 750 billion Suns and an optical diameter of 175 thousand light years. It has a small (5 billion solar mass) companion galaxy (ESO 495-G017, mag. 15, 7′ NW).

NGC 2627

Constellation	Object type	RA, Dec	Approx. transit date at local midnight	Distance
Pyxis	Open cluster	08h 37.2m, –29° 57′	February 8	6 thousand light years
Age	**Apparent size**	**Magnitude**	**Sky Atlas 2000.0 chart**	**Herald–Bobroff chart**
4 hundred million years	11′	8.4	20	C69

It has a diameter of about 20 light years and was discovered in 1793 by William Herschel.

NGC 2632 (M 44)

Constellation	Object type	RA, Dec	Approx. transit date at local midnight	Distance
Cancer	Open cluster	08h 40.0m, +19° 40′	February 9	6 hundred light years
Age	**Apparent size**	**Magnitude**	**Sky Atlas 2000.0 chart**	**Herald–Bobroff chart**
7 hundred million years	1.6°	3.1	6	C32

Nicknamed the "Beehive Cluster" or "Praesepe" (which means manger in Latin, its common-use anglicized pronunciation being pree-SEE-pee). Recent work suggests that M 44 is actually two merging clusters. The total cluster mass is about 7 hundred solar masses. M 44 and the Hyades (60° W) may be part of a single, moving group (some suggest the two formed from a single gaseous cloud). M 44's size of 1.6° corresponds to a diameter of about 16 light years.

NGC 2655

Constellation	Object type	RA, Dec	Approx. transit date at local midnight	Distance
Camelopardalis	Barred lenticular galaxy	08h 55.6m, +78° 13'	February 13	80 million light years
Age	**Apparent size**	**Magnitude**	**Sky Atlas 2000.0 chart**	**Herald–Bobroff chart**
	4.9' × 4.1'	10.1	2	C15

It has an optical diameter of a little over a hundred thousand light years. It rotates about an axis that is inclined at 34° from our line-of-sight. The nucleus has a mass of about 10 billion Suns within a diameter of 2 thousand light years of the center and its rotation appears to be decoupled from the motion of the rest of the galaxy. The nucleus has strong emissions associated with photoionization intermediate between that of Seyfert (see NGC 3227) and LINER galaxies (see M 81/NGC 3031).

NGC 2681

Constellation	Object type	RA, Dec	Approx. transit date at local midnight	Distance
Ursa Major	Barred lenticular galaxy	08h 53.5m, +51° 19'	February 13	45 million light years
Age	**Apparent size**	**Magnitude**	**Sky Atlas 2000.0 chart**	**Herald–Bobroff chart**
	3.7' × 3.7'	10.3	1	C14

This galaxy has an optical diameter of about 50 thousand light years and is nearly face-on. The nucleus of this galaxy is thought to harbor a central black hole with a mass of at most 60 million Suns, and probably close to 8 million Suns. This central black hole is accreting gas and stars, resulting in photoionization of surrounding gas and a nuclear emission region (that is less than 20 light years in diameter), so that this is a LINER galaxy ("low-ionization nuclear emission region" – see M 81/NGC 3031 for further explanation). The central region of this galaxy (within 2 thousand light years of the nucleus) is thought to have undergone a short burst of star formation about 1 billion years ago (like that which is currently occurring in M 82), perhaps due to a merger with another galaxy.

NGC 2682 (M 67)

Constellation	Object type	RA, Dec	Approx. transit date at local midnight	Distance
Cancer	Open cluster	08h 50.8m, +11° 49'	February 11	3 thousand light years
Age	**Apparent size**	**Magnitude**	**Sky Atlas 2000.0 chart**	**Herald–Bobroff chart**
4 billion years	30'	6.9	12	C50

This is one of the oldest open clusters known. Open clusters that orbit closer in to the Galactic center are disrupted more readily, so that almost all old open clusters have orbits that stay at least as far from the Galactic center as we are located (i.e. at radii > 25 thousand light years or so). Professional telescopes find about 5 hundred members (down to magnitude 17) in this cluster. Its size of 30' corresponds to a diameter of about 25 light years.

NGC 2683

Constellation	Object type	RA, Dec	Approx. transit date at local midnight	Distance
Lynx	Spiral galaxy	08h 52.7m, +33° 25′	February 12	30 million light years
Age	**Apparent size**	**Magnitude**	**Sky Atlas 2000.0 chart**	**Herald–Bobroff chart**
	8.8′ × 2.5′	9.8	6	C32

It has an optical diameter of about 80 thousand light years. Professional telescopic studies indicate that it probably contains over 3 hundred globular clusters. It is nearly edge-on to us, with an inclination angle (the angle between its rotation axis and our line-of-sight) of 82°.

NGC 2742

Constellation	Object type	RA, Dec	Approx. transit date at local midnight	Distance
Ursa Major	Spiral galaxy	09h 07.6m, +60° 29′	February 16	65 million light years
Age	**Apparent size**	**Magnitude**	**Sky Atlas 2000.0 chart**	**Herald–Bobroff chart**
	3.0′ × 1.5′	11.4	2	C14

It has an optical diameter of about 60 thousand light years. There is a large central concentration of mass (about 10 billion Suns within the inner thousand light years, which has an apparent radius of 2″). Most of the disk of this galaxy rotates at about one hundred km/s (the rotation axis is inclined at an angle of about 60° from our line-of-sight).

NGC 2768

Constellation	Object type	RA, Dec	Approx. transit date at local midnight	Distance
Ursa Major	Elliptical galaxy	09h 11.6m, +60° 02′	February 17	65 million light years
Age	**Apparent size**	**Magnitude**	**Sky Atlas 2000.0 chart**	**Herald–Bobroff chart**
	8.2′ × 5.3′	9.9	2	C14

The diameter of this galaxy is about 150 thousand light years. It has a low-ionization nuclear emission region ("LINER" – see M 81/NGC 3031). It forms a gravitationally bound group of five galaxies with the spiral galaxies NGC 2742 (40′ NW – see NGC 2742), NGC 2726 (50′ W, mag. 13), NGC 2654 (2° 48′ W, mag. 11.8) and UGC 4549 (3° 40′ WSW, mag. 13), which all lie within several million light years of each other.

NGC 2775

Constellation	Object type	RA, Dec	Approx. transit date at local midnight	Distance
Cancer	Spiral galaxy	09h 10.3m, +07° 02′	February 17	60 million light years
Age	**Apparent size**	**Magnitude**	**Sky Atlas 2000.0 chart**	**Herald–Bobroff chart**
	4.5′ × 3.6′	10.1	12	C50

The total luminous mass of this galaxy is about 150 billion Suns, about 40% of which is in the disk and the rest is in the central bulge. Dark matter, largely in the halo of this galaxy, adds at most another 30 billion Suns to the mass of this galaxy. The galaxy rotates about an axis inclined at 44° from our line-of-sight. Its position angle is 155° (i.e. the major axis of its elliptical shape is aligned 155° counterclockwise from north, which is a SE direction). It has an optical diameter of about 80 thousand light years.

NGC 2782

Constellation	Object type	RA, Dec	Approx. transit date at local midnight	Distance
Lynx	Barred spiral galaxy	09h 14.1m, +40° 07′	February 17	120 million light years
Age	**Apparent size**	**Magnitude**	**Sky Atlas 2000.0 chart**	**Herald–Bobroff chart**
	3.7′ × 2.4′	11.6	6	C32

This galaxy has an optical diameter of about 130 thousand light years. Intense star formation (called a "starburst") is occurring within a 20″ diameter region at the center of this galaxy, creating supernovae and stellar winds that drive a powerful outflow from this region. Professional telescopic studies find a gas-rich bar (of radius 7.5″) exists in the nucleus of this galaxy, which is thought to be funneling gas into the starburst region. The starburst and nuclear bar are expected to die away within 5 hundred million years. This galaxy underwent a merger in the recent past (some tens of millions of years ago), which is the cause of the two "tails" to this galaxy that appear in professional telescopic images.

NGC 2787

Constellation	Object type	RA, Dec	Approx. transit date at local midnight	Distance
Ursa Major	Barred lenticular galaxy	09h 19.3m, +69° 12′	February 19	40 million light years
Age	**Apparent size**	**Magnitude**	**Sky Atlas 2000.0 chart**	**Herald–Bobroff chart**
	3.1′ × 1.8′	10.8	2	C14

This galaxy has an optical diameter of about 40 thousand light years. It rotates about an axis inclined at 50° from our line-of-sight. It is a LINER (low-ionization nuclear emission region) galaxy, with emission in its nucleus thought to be associated with accretion onto a central supermassive black hole (having a mass of about a hundred million Suns for this galaxy). This galaxy has an unusually high value of mass-to-luminosity (a value of 50 in solar mass and solar luminosity units), which is thought to be due to this galaxy having a normal mass but a very low luminosity (about that of a dwarf galaxy).

NGC 2811

Constellation	Object type	RA, Dec	Approx. transit date at local midnight	Distance
Hydra	Barred spiral galaxy	09h 16.2m, −16° 19′	February 18	110 million light years
Age	**Apparent size**	**Magnitude**	**Sky Atlas 2000.0 chart**	**Herald–Bobroff chart**
	2.2′ × 0.7′	11.3	12	C50

It has an optical diameter of 70 thousand light years. It is nearly edge-on, with an inclination angle of about 70° (an angle of 90° would be edge-on).

NGC 2841

Constellation	Object type	RA, Dec	Approx. transit date at local midnight	Distance
Ursa Major	Spiral galaxy	09h 22.0m, +50° 59′	February 19	50 million light years
Age	**Apparent size**	**Magnitude**	**Sky Atlas 2000.0 chart**	**Herald–Bobroff chart**
	7.7′ × 3.6′	9.2	2	C14

This galaxy has a weak LINER (low-ionization nuclear emission region – see M 81/NGC 3031 for explanation) in its nucleus. In addition, its nucleus has a different chemical composition from the rest of the galaxy. The galaxy rotates about an angle that is inclined at 68° from our line-of-sight. It has a relatively high rotation rate – about 270 km/s throughout most of the disk. It also has a giant reflection nebula that covers an area of 4 thousand × 7 thousand light years slightly above the disk on the near side of the galaxy. Because of the galaxy's inclination to our line-of-sight, light from the central bulge in the galaxy reflects off the dust in this nebula, making this huge dust cloud visible in professional telescopic images. The nebula is very dense – light is dimmed by 13 magnitudes for every kiloparsec (3.26 thousand light years) that it travels through this nebula (compare this to the average 2 magnitudes of dimming that occurs over a similar distance in the central plane of our Galaxy). This is one of fewer than ten galaxies that has had four or more recorded supernovae.

NGC 2859

Constellation	Object type	RA, Dec	Approx. transit date at local midnight	Distance
Leo Minor	Barred lenticular galaxy	09h 24.3m, +34° 31′	February 20	80 million light years
Age	**Apparent size**	**Magnitude**	**Sky Atlas 2000.0 chart**	**Herald–Bobroff chart**
	4.0′ × 3.6′	10.9	6	C32

In professional telescopic studies this galaxy has a detached outer ring around it, making it unusual, since only about 10% of galaxies have this feature. However, outer rings are common with barred lenticular galaxies (about 2/3 of them have an outer ring). This may be due to the more dynamically evolved state of lenticular galaxies compared to other galaxies, since outer rings are thought to occur relatively late in a galaxy's evolution. Outer rings also require a galaxy to avoid tidal interactions with other galaxies (since such interactions destroy these rings). As a result, galaxies that are part of groups rarely have outer rings. This galaxy has an optical diameter of about 90 thousand light years.

NGC 2903

Constellation	Object type	RA, Dec	Approx. transit date at local midnight	Distance
Leo	Barred spiral galaxy	09h 32.2m, +21° 30′	February 22	30 million light years
Age	Apparent size	Magnitude	Sky Atlas 2000.0 chart	Herald–Bobroff chart
	12.0′ × 5.6′	9.0	6	C32

Star-forming regions in its spiral arms that are bright ionized hydrogen emission (HII) regions give the disk a splotchy appearance that is visible in large amateur telescopes. This is also a "starburst" galaxy where rapid star formation is occurring in its nucleus. These star-formation regions appear as splotchy "hotspots" in professional telescopic images of the nucleus because gas and dust in the nucleus block the view. This galaxy has an optical diameter of about 1 hundred thousand light years.

NGC 2950

Constellation	Object type	RA, Dec	Approx. transit date at local midnight	Distance
Ursa Major	Barred lenticular galaxy	09h 42.6m, +58° 51′	February 25	75 million light years
Age	Apparent size	Magnitude	Sky Atlas 2000.0 chart	Herald–Bobroff chart
	2.7′ × 1.8′	10.9	2	C14

Professional telescopic studies show that the disk of this galaxy gives off 48% of the total light from the galaxy, while the nucleus accounts for 29% of the light and the bar the remaining 22%. It rotates about an axis that is inclined to our line-of-sight by about 48° and it has a position angle of 144° (i.e. its major axis lies 144° counterclockwise from N).

NGC 2964

Constellation	Object type	RA, Dec	Approx. transit date at local midnight	Distance
Leo	Barred spiral galaxy	09h 42.9m, +31° 51′	February 25	70 million light years
Age	Apparent size	Magnitude	Sky Atlas 2000.0 chart	Herald–Bobroff chart
	3.0′ × 1.7′	11.3	6	C32

It forms a noninteracting triplet with NGC 2968 (irregular galaxy, mag. 12, 6′ NNE) and NGC 2970 (elliptical galaxy, mag. 13.6, 11′ NE). NGC 2964 has an optical diameter of 60 thousand light years, while NGC 2968 has an optical diameter of about 45 thousand light years. NGC 2964 rotates about an axis inclined at about 55° from our line-of-sight (and has a position angle of 96° i.e. its major axis is aligned nearly E–W), while NGC 2968 has an inclination angle of 40° and a position angle of 44° (i.e. elongated NE–SW).

NGC 2974

Constellation	Object type	RA, Dec	Approx. transit date at local midnight	Distance
Sextans	Elliptical galaxy	09h 42.6m, –03° 42′	February 25	90 million light years
Age	**Apparent size**	**Magnitude**	**Sky Atlas 2000.0 chart**	**Herald– Bobroff chart**
	3.4′ × 2.1′	10.9	13	C50

This galaxy has an unusually high rotation rate for an elliptical galaxy (up to 250 km/s). Despite being classified as an elliptical galaxy, dynamical modeling suggests that it actually has an embedded stellar disk (making it a lenticular galaxy). However, disks in elliptical galaxies are difficult to discern unless the galaxy is nearly edge-on. This galaxy is inclined at an angle of about 60° from our line-of-sight (edge-on would have an inclination angle of 90°), so that its disk is not easily detected in professional telescopic studies. The disk is thought to account for about 7% of the light from this galaxy. The luminous mass of this galaxy is several hundred billion Suns, and dark matter in its halo more than doubles this value. A low-ionization nuclear emission region ("LINER") is present in the region out to a radius of 30″ from the center (see M 81/NGC 3031 for explanation).

NGC 2976

Constellation	Object type	RA, Dec	Approx. transit date at local midnight	Distance
Ursa Major	Peculiar spiral galaxy	09h 47.2m, +67° 55′	February 26	13 million light years
Age	**Apparent size**	**Magnitude**	**Sky Atlas 2000.0 chart**	**Herald– Bobroff chart**
	6.2′ × 3.1′	10.8	2	C14

It is part of the M 81 group of gravitationally bound galaxies (see M 81/NGC 3031). Indeed, professional telescopic studies find a bridge of neutral hydrogen (HI) gas that is 1/4 million light years long and connects NGC 2976 to M 81. This bridge has a mass of about 20 billion Suns and is the result of gravitational interactions between these galaxies.

NGC 2985

Constellation	Object type	RA, Dec	Approx. transit date at local midnight	Distance
Ursa Major	Spiral galaxy	09h 50.4m, +72° 17′	February 27	70 million light years
Age	**Apparent size**	**Magnitude**	**Sky Atlas 2000.0 chart**	**Herald– Bobroff chart**
	4.6′ × 3.4′	10.4	2	C14

This galaxy has a low-ionization nuclear emission region (LINER – see M 81/NGC 3031 for explanation) as well as many ionized hydrogen (HII) star-formation regions, particularly within a radius of 5–30″ of the galaxy's center. These regions give the galaxy a mottled appearance outside its nucleus in large amateur telescopes. The galaxy forms a noninteracting (gravitationally bound) pair with NGC 3027 (barred spiral, 25′ E, mag. 11.8) and rotates about an axis that is inclined by an angle of about 40° from our line-of-sight.

NGC 3003

Constellation	Object type	RA, Dec	Approx. transit date at local midnight	Distance
Leo Minor	Barred spiral galaxy	09h 48.6m, +33° 25'	February 26	80 million light years
Age	**Apparent size**	**Magnitude**	**Sky Atlas 2000.0 chart**	**Herald– Bobroff chart**
	5.7' × 1.4'	11.9	6	C32

It has a mass of about 50 billion Suns and is nearly edge-on. Active star formation is occurring in this galaxy with many ionized hydrogen star-forming regions (i.e. HII regions) some of which are visible as bright knots in large amateur telescopes. Nearby NGC 3021 (31' E) is thought to be about 20 million light years closer than NGC 3003.

NGC 3031 (M 81)

Constellation	Object type	RA, Dec	Approx. transit date at local midnight	Distance
Ursa Major	Spiral galaxy	09h 55.6m, +69° 04'	February 28	13 million light years
Age	**Apparent size**	**Magnitude**	**Sky Atlas 2000.0 chart**	**Herald– Bobroff chart**
	24.9' × 11.5'	6.9	2	C13

M 81 is a LINER (low-ionization nuclear emission region) galaxy, which is a low-luminosity class of "active galactic nuclei" (AGN). The mechanism for LINERs remains uncertain but in some galaxies it may be due to a supermassive black hole in the nucleus that is accreting gas and stars, resulting in photoionization of surrounding gas. (Indeed, some suggest that LINER galaxies represent an evolutionary stage between quasars and ordinary galaxies.) Alternatively, some LINERs may instead be caused by intense star-formation activity in the nucleus (a "starburst"). M 81 is thought to be in the former class (a LINER whose emission is associated with a central black hole), with its supermassive black hole estimated to contain about 60 million solar masses. About one third of all galaxies are LINERs. M 81 is the namesake member of the M 81 group of at least 11 gravitationally bound galaxies that includes NGC 2403 (14° W), NGC 2976 (1° 23' SW), IC 2574 (3° E), NGC 4236 (12° E), in addition to nearby M 82/NGC 3034 and NGC 3077 with which M 81 has had strong past interactions (see NGC 3077 and M 82/NGC 3034). M 81 has an optical diameter of about 95 thousand light years and a mass of about 50 billion Suns.

NGC 3034 (M 82)

Constellation	Object type	RA, Dec	Approx. transit date at local midnight	Distance
Ursa Major	Irregular galaxy	09h 55.9m, +69° 41'	February 28	13 million light years
Age	**Apparent size**	**Magnitude**	**Sky Atlas 2000.0 chart**	**Herald–Bobroff chart**
	10.5' × 5.1'	8.4	2	C14

This is the prototypical starburst galaxy (in which intense star formation is occurring in its central region). Indeed, three times as many stars are forming every year in the center of this galaxy (within a radius of about 0.5' of this galaxy's center) as form in the entire Milky Way in one year. Supernovae occur in this starburst region about once a decade (which is several times the rate for the entire Milky Way). These supernovae blow material out of the center of this galaxy in a superwind (moving at 5–8 hundred km/s) that forms two jets perpendicular to the plane of the galaxy. These jets (which pick up material on their way out, possibly by turbulent shear layer mixing and by evaporating nearby gas in the galaxy), are believed to be slamming into gas in a halo outside the galaxy's disk. This halo gas is thought to be left over from earlier gravitational interactions with nearby M 81. Material in the superwind jets is thought to be moving faster than the escape velocity of the galaxy and so will become intergalactic material. M 82 is part of the M 81 group of gravitationally bound galaxies (see M 81/NGC 3031) and is thought to have had strong interactions with M 81 over the last several hundred million years that have triggered the starburst in M 82. Associated with the bright star-forming regions, over one hundred "super star clusters" are known in M 82, thought to be newly minted globular clusters containing millions of Suns each. The optical diameter of M 82 is about 40 thousand light years.

NGC 3077

Constellation	Object type	RA, Dec	Approx. transit date at local midnight	Distance
Ursa Major	Irregular galaxy	10h 03.3m, +68° 44'	March 2	13 million light years
Age	**Apparent size**	**Magnitude**	**Sky Atlas 2000.0 chart**	**Herald–Bobroff chart**
	5.2' × 4.7'	9.9	2	C14

This galaxy has an optical diameter of about 20 thousand light years. It is considered a dwarf galaxy and belongs to the M 81 group of gravitationally bound galaxies (see M 81/NGC 3031). It has interacted with M 81 (the strongest such interaction having occurred several hundred million years ago), and professional telescopic studies show that long streamers of neutral hydrogen gas bridge the two galaxies. Like nearby M 82, NGC 3077 is a starburst galaxy in which intense star formation is occurring in its nucleus. The starburst in NGC 3077 is thought to be fueled by gas falling back into the galaxy after having been torn out in an earlier interaction with M 81.

NGC 3079

Constellation	Object type	RA, Dec	Approx. transit date at local midnight	Distance
Ursa Major	Barred spiral galaxy	10h 02.0m, +55° 41'	March 2	55 million light years
Age	**Apparent size**	**Magnitude**	**Sky Atlas 2000.0 chart**	**Herald–Bobroff chart**
	8.1' × 1.3'	10.9	2	C13

The nucleus of this galaxy contains the most luminous known water vapor maser. Maser stands for "microwave amplification by stimulated emission of radiation", the physics of which is the microwave equivalent of a laser (except that lasers are usually designed to produce a beam, while astronomical masers yield emission that radiates from a roughly spherical region). The nucleus of this galaxy appears to be both an AGN (or "active galactic nucleus", in which energetic emission is caused by accretion onto a massive central object) and a starburst (in which intense star formation is occurring). A superbubble driven by either the AGN or starburst has created jets flowing out of the nucleus perpendicular to the disk. This galaxy has an optical diameter of about 130 thousand light years.

NGC 3115

Constellation	Object type	RA, Dec	Approx. transit date at local midnight	Distance
Sextans	Lenticular galaxy	10h 05.2m, –07° 43'	March 2	30 million light years
Age	**Apparent size**	**Magnitude**	**Sky Atlas 2000.0 chart**	**Herald–Bobroff chart**
	7.3' × 3.4'	8.9	13	C49

Nicknamed the "Spindle Galaxy", although NGC 5866 also has this nickname. NGC 3115 is believed to harbor a supermassive black hole in its nucleus, with a mass of about a billion Suns. However, it is not an active galactic nucleus (i.e. intense emission is not occurring from the nucleus), so that this black hole is not being fed well. The disk in this lenticular galaxy accounts for only about 6% of its total luminosity. The galaxy is nearly edge-on (inclination angle of 86°) and its apparent size corresponds to an optical diameter of about 60 thousand light years, although a stellar halo extends to about 1 hundred thousand light years in professional telescopic studies.

NGC 3147

Constellation	Object type	RA, Dec	Approx. transit date at local midnight	Distance
Draco	Spiral galaxy	10h 16.9m, +73° 24'	March 5	130 million light years
Age	**Apparent size**	**Magnitude**	**Sky Atlas 2000.0 chart**	**Herald–Bobroff chart**
	4.3' × 3.7'	10.6	2	C13

This galaxy has an active galactic nucleus (AGN, where a supermassive object in the galaxy's center accumulates nearby gas resulting in strong emission from the nucleus) of a Seyfert type. Very active star formation is occurring in a ring that lies at a radius of 7–8 thousand light years (about 10 arcseconds) from the center of the galaxy. It is one of fewer than about a hundred galaxies that have had two or more recorded supernovae. It is nearly face-on (with an inclination angle of 27°) and has an optical diameter of about 160 thousand light years.

NGC 3166

Constellation	Object type	RA, Dec	Approx. transit date at local midnight	Distance
Sextans	Barred lenticular galaxy	10h 13.8m, +03° 26′	March 5	70 million light years
Age	**Apparent size**	**Magnitude**	**Sky Atlas 2000.0 chart**	**Herald–Bobroff chart**
	4.8′ × 1.9′	10.4	13	C49

It is nearly edge-on (inclination angle of 73°). It is the namesake member of the NGC 3166 group of five gravitationally bound galaxies that includes nearby NGC 3169 (spiral galaxy, 6′ ENE, mag. 10.2 – see NGC 3169), the dimmer NGC 3156 (spiral galaxy, 23.5′ SW, mag. 12.3), and the much dimmer NGC 3165 (lenticular galaxy, 5′ SW, mag. 13.9). NGC 3166 is a LINER (low-ionization nuclear emission region) galaxy (see M 81/NGC 3031 for explanation).

NGC 3169

Constellation	Object type	RA, Dec	Approx. transit date at local midnight	Distance
Sextans	Spiral galaxy	10h 14.2m, +03° 28′	March 5	70 million light years
Age	**Apparent size**	**Magnitude**	**Sky Atlas 2000.0 chart**	**Herald–Bobroff chart**
	4.2′ × 2.9′	10.2	13	C49

It has a LINER (low-ionization nuclear emission region – see M 81/NGC 3031 for further explanation). It is part of the NGC 3166 group of gravitationally bound galaxies. It is thought to be interacting (i.e. changing shape or mass due to gravitational interaction) with NGC 3166 (see NGC 3166). It rotates about an axis that is inclined at about 55° from our line-of-sight.

NGC 3184

Constellation	Object type	RA, Dec	Approx. transit date at local midnight	Distance
Ursa Major	Barred spiral galaxy	10h 18.3m, +41° 25′	March 6	30 million light years
Age	**Apparent size**	**Magnitude**	**Sky Atlas 2000.0 chart**	**Herald–Bobroff chart**
	7.6′ × 7.4′	9.8	6	C31

This galaxy has had a remarkable four supernovae observed in it (with two occurring in 1921). Fewer than 10 galaxies share this accomplishment and only two galaxies have had more observed supernovae (NGC 6946 and M 83, which have had seven and six, respectively). It is nearly face-on (with an inclination angle of about 20°). Several star-forming, ionized hydrogen (HII) regions are visible as knots in its disk in large amateur telescopes. It has an optical diameter of about 65 thousand light years. It does not have an active galactic nucleus (see NGC 3079 for explanation).

NGC 3190

Constellation	Object type	RA, Dec	Approx. transit date at local midnight	Distance
Leo	Spiral galaxy	10h 18.1m, +21° 50'	March 6	50 million light years
Age	Apparent size	Magnitude	Sky Atlas 2000.0 chart	Herald–Bobroff chart
	4.0' × 1.5'	11.1	6	C31

It is nearly edge-on (it rotates about an angle inclined at 70° from our line-of-sight). It has an optical diameter of about 60 thousand light years, and has a low-ionization nuclear emission region (LINER – see M 81/NGC 3031 for explanation). It is the name-sake member of the NGC 3190 group of 13 gravitationally bound galaxies. The 10 NGC members of this group include NGC 3162 (barred spiral galaxy, 1° 24' NW, mag. 11.6), NGC 3185 (barred spiral galaxy, 11' SW, mag. 12.2), NGC 3193 (6' NE – see NGC 3193), NGC 3177 (spiral galaxy, 48' SSW, mag. 12.3), NGC 3187 (barred spiral, 5' WNW, mag. 13.4), NGC 3213 (spiral, 2° 18' SSE, mag. 13.5), NGC 3226 (2° 18' SSE – see NGC 3226), NGC 3227 (2° 20' SSE – see NGC 3227), NGC 3287 (barred spiral, 3° 53' E, mag. 12.3) and NGC 3301 (lenticular galaxy, 4° 22' E, mag. 11.4).

NGC 3193

Constellation	Object type	RA, Dec	Approx. transit date at local midnight	Distance
Leo	Elliptical galaxy	10h 18.4 m, +21° 54'	March 6	50 million light years
Age	Apparent size	Magnitude	Sky Atlas 2000.0 chart	Herald–Bobroff chart
	2.9' × 2.8'	10.9	6	C31

It is part of the NGC 3190 galaxy group (see NGC 3190). It has an optical diameter of about 40 thousand light years.

NGC 3198

Constellation	Object type	RA, Dec	Approx. transit date at local midnight	Distance
Ursa Major	Barred spiral galaxy	10h 19.0 m, +45° 33'	March 6	45 million light years
Age	Apparent size	Magnitude	Sky Atlas 2000.0 chart	Herald–Bobroff chart
	8.1' × 3.0'	10.3	6	C31

It is nearly edge-on (with an inclination angle of about 70°). Professional telescopic studies show this galaxy has little or no bulge (the bulge is the central spherical region in disk galaxies). Most of the disk that makes up the luminous part of this galaxy rotates at about 150 km/s. About half of the mass of this galaxy resides in a spherical dark matter halo which is thought to be very extended and of such low density that it has little effect on the dynamics of the disk. It is one of fewer than about a hundred galaxies that have had two or more recorded supernovae.

NGC 3226

Constellation	Object type	RA, Dec	Approx. transit date at local midnight	Distance
Leo	Elliptical galaxy	10h 23.4 m, +19° 54'	March 7	50 million light years
Age	**Apparent size**	**Magnitude**	**Sky Atlas 2000.0 chart**	**Herald–Bobroff chart**
	2.5' × 2.2'	11.4	6	C31

It is part of the NGC 3190 galaxy group (see NGC 3190). It is a dwarf galaxy and is thought to harbor a low-luminosity active galactic nucleus (AGN – see NGC 3079 for explanation) which is a LINER (low-ionization nuclear emission region – see M 81/NGC 3031). It is interacting with nearby NGC 3227 (which is 2.5' SE – see NGC 3227).

NGC 3227

Constellation	Object type	RA, Dec	Approx. transit date at local midnight	Distance
Leo	Barred spiral galaxy	10h 23.5 m, +19° 52'	March 7	50 million light years
Age	**Apparent size**	**Magnitude**	**Sky Atlas 2000.0 chart**	**Herald–Bobroff chart**
	6.6' × 5.0'	10.3	6	C31

It is part of the NGC 3190 galaxy group (see NGC 3190). This is a Seyfert galaxy, one of twelve such galaxies first identified by Seyfert himself, in which emission from the nucleus is thought to occur due to accretion of matter onto a massive central black hole. About 20 million solar masses are thought to be present within a radius of 80 light years (0.3") of the center of this galaxy. Molecular gas close to the nucleus is thought to be partly obscuring the nucleus, reducing the intensity of nuclear emission that we see.

NGC 3242

Constellation	Object type	RA, Dec	Approx. transit date at local midnight	Distance
Hydra	Planetary nebula	10h 24.8 m, −18° 39'	March 7	2 thousand light years
Age	**Apparent size**	**Magnitude**	**Sky Atlas 2000.0 chart**	**Herald–Bobroff chart**
	40"	8.6	13	C67

Nicknamed the "Ghost of Jupiter" since it is about the same apparent size as Jupiter. The central star (mag. 12.1), visible in amateur telescopes, has a surface temperature of about 60 thousand K and has a mass less than half that of our Sun. The bright inner ring (with a diameter of about 21") visible in large amateur telescopes is expanding at about 25 km/s and has an actual diameter of about 0.4 light years. This ring is thought to be part of a roughly cylindrical shell with the axis of the cylinder inclined at 52° from our line-of-sight, making the cross-section of the cylinder that we see as a ring have an elliptical shape with major axis having a position angle of 56°. In professional telescopes, the NW and SW edges of the inner ring have fast, low-ionization emission regions ("FLIERS" – see NGC 7662), and the nebula extends out in a patchy halo to a diameter of over 3', only the inner portions of which are visible in amateur telescopes.

NGC 3245

Constellation	Object type	RA, Dec	Approx. transit date at local midnight	Distance
Leo Minor	Lenticular galaxy	10h 27.3 m, +28° 30′	March 8	75 million light years
Age	Apparent size	Magnitude	Sky Atlas 2000.0 chart	Herald–Bobroff chart
	3.5′ × 2.4′	10.8	6	C31

This galaxy has a circumnuclear disk (with radius 1″) that is photoionized from matter accreting onto a central black hole containing 2 hundred million solar masses. This emission is responsible for the LINER (low-ionization nuclear emission region) classification of this galaxy. It is part of the NGC 3254 group of galaxies (see NGC 3277).

NGC 3277

Constellation	Object type	RA, Dec	Approx. transit date at local midnight	Distance
Leo Minor	Spiral galaxy	10h 32.9 m, +28° 31′	March 9	75 million light years
Age	Apparent size	Magnitude	Sky Atlas 2000.0 chart	Herald–Bobroff chart
	2.1′ × 1.8′	11.7	6	C31

It has an optical diameter of about 50 thousand light years. It is nearly face-on (having an inclination angle of about 27°, meaning that it rotates about an axis that is inclined from our line-of-sight by 27°). It is part of the NGC 3254 group of five gravitationally bound galaxies, that includes NGC 3245 (1° 14′ W – see NGC 3245) and NGC 3254 (spiral galaxy, 1° 16′ NW, mag. 11.7), as well as the much dimmer NGC 3265 and NGC 3245A.

NGC 3294

Constellation	Object type	RA, Dec	Approx. transit date at local midnight	Distance
Leo Minor	Spiral galaxy	10h 36.3 m, +37° 19′	March 10	90 million light years
Age	Apparent size	Magnitude	Sky Atlas 2000.0 chart	Herald–Bobroff chart
	3.4′ × 1.8′	11.8	6	C31

It has an optical diameter of about 90 thousand light years. It rotates about an axis inclined from our line-of-sight by about 60° (which is its "inclination angle"). The major axis of the ellipse that results by viewing its circular disk at this inclination angle is along a line that is 122° counterclockwise from north (which is its "position angle"). It is one of fewer than a hundred galaxies that have had two or more recorded super-novae (which occurred two years apart – in 1990 and 1992).

NGC 3310

Constellation	Object type	RA, Dec	Approx. transit date at local midnight	Distance
Ursa Major	Peculiar barred spiral galaxy	10h 38.8 m, +53° 30′	March 11	40 million light years
Age	**Apparent size**	**Magnitude**	**Sky Atlas 2000.0 chart**	**Herald–Bobroff chart**
	2.8′ × 2.6′	10.8	2	C13

This galaxy has a number of interesting features in professional telescopic studies, including
- A 3 thousand light year (16″) diameter circumnuclear ring of intense star formation (i.e. a "starburst"), which has been active for about one hundred million years.
- A "bow and arrow" shape on the western side. The bow is at a radius of 30″ and thought to contain older stars which are perhaps debris from the disk of a cannibalized galaxy, while the "arrow" points NW extending from about 20″ to 50″ from the galaxy center and is made of hot, young stars the same age as the starburst ring just mentioned.
- A 1.5″ offset between the stellar nucleus and the dynamical center of the galaxy.

These and other features suggest that this galaxy underwent a merger with another galaxy in the not too distant past. The galaxy has a total mass of several tens of billions of Suns.

NGC 3344

Constellation	Object type	RA, Dec	Approx. transit date at local midnight	Distance
Leo Minor	Barred spiral galaxy	10h 43.5 m, +24° 55′	March 12	20 million light years
Age	**Apparent size**	**Magnitude**	**Sky Atlas 2000.0 chart**	**Herald–Bobroff chart**
	7.1′ × 6.8′	9.9	6	C31

Professional telescopes show this galaxy has an inner ring (0.5′ in diameter) and an outer ring (3′ in diameter) around it, both of which are star-forming regions. Inner rings are a common feature of spiral galaxies (about 60% of all spiral galaxies have them), but outer rings are rarer (with only 10% of galaxies having an outer ring or pseudo-ring). Outer rings are even more uncommon in "late-type" spirals like this galaxy (where the spiral arms are loosely "wound" and there is only a small central bulge compared to the extended disk), and it has been hypothesized that this galaxy's outer ring may be due to a past merger. This galaxy is nearly face-on (with an inclination angle of about 25°) and has an optical diameter of 40 thousand light years.

NGC 3351 (M 95)

Constellation	Object type	RA, Dec	Approx. transit date at local midnight	Distance
Leo	Barred spiral galaxy	10h 44.0 m, +11° 42′	March 12	35 million light years
Age	**Apparent size**	**Magnitude**	**Sky Atlas 2000.0 chart**	**Herald–Bobroff chart**
	7.3′ × 4.4′	9.7	13	C49/D36

It is part of the M 96 (NGC 3368) galaxy group – see M 96/NGC 3368. Professional telescopes find that M 95 has an inner ring (7″ in radius) and an outer ring (1′ in diameter), both being ionized hydrogen (HII) star-forming regions and due to orbital resonant interactions with this galaxy's bar. M 95 is a nearly average galaxy in size and mass, having a mass of about 50 billion Suns and an optical diameter of about 70 thousand light years.

NGC 3368 (M 96)

Constellation	Object type	RA, Dec	Approx. transit date at local midnight	Distance
Leo	Barred spiral galaxy	10h 46.8 m, +11° 49′	March 13	35 million light years
Age	**Apparent size**	**Magnitude**	**Sky Atlas 2000.0 chart**	**Herald–Bobroff chart**
	7.8′ × 5.2′	9.3	13	C49/D36

It is the namesake member of the M 96 (NGC 3368) group of 12 galaxies that includes NGC 3299, NGC 3351 (M 95), NGC 3377 (see NGC 3377), M 105 (NGC 3379), NGC 3384 (see NGC 3384), NGC 3412 (see NGC 3412), and NGC 3489 (see NGC 3489). M 96 has a mass of about 80 billion Suns, an optical diameter of about 80 thousand light years and is a LINER (see M 81/NGC 3031).

NGC 3377

Constellation	Object type	RA, Dec	Approx. transit date at local midnight	Distance
Leo	Elliptical galaxy	10h 47.7 m, +13° 59′	March 13	35 million light years
Age	**Apparent size**	**Magnitude**	**Sky Atlas 2000.0 chart**	**Herald–Bobroff chart**
	4.3′ × 2.6′	10.4	13	C31/D36

The center of this galaxy is believed to harbor a supermassive black hole containing about 2 hundred million solar masses. This galaxy is part of the M 96 (NGC 3368) group of galaxies that includes nine gravitationally bound galaxies (see M 96/NGC 3368). The galaxy is a flattened ellipsoid in shape (elliptical galaxies don't get much flatter than this one) with a minor/major axis ratio of 0.5, viewed nearly edge-on.

NGC 3379 (M 105)

Constellation	Object type	RA, Dec	Approx. transit date at local midnight	Distance
Leo	Elliptical galaxy	10h 47.8 m, +12° 35′	March 13	35 million light years
Age	**Apparent size**	**Magnitude**	**Sky Atlas 2000.0 chart**	**Herald–Bobroff chart**
	5.3′ × 4.8′	9.3	13	C31/D36

It is part of the M 96 (NGC 3368) group of galaxies (see M 96/NGC 3368). It has a mass of about one hundred billion Suns. It contains a supermassive central dark mass (thought to be a black hole) with a mass of one or 2 hundred million Suns. NGC 3389 (spiral galaxy, 10′ ESE, mag. 11.9, with a mass similar to M 105) is nearby, although it is actually a background object and is instead about twice as far away and part of the NGC 3338 group of galaxies that includes NGC 3338, NGC 3389 andNGC 3346.

NGC 3384

Constellation	Object type	RA, Dec	Approx. transit date at local midnight	Distance
Leo	Barred lenticular galaxy	10h 48.3 m, +12° 38′	March 13	35 million light years
Age	**Apparent size**	**Magnitude**	**Sky Atlas 2000.0 chart**	**Herald–Bobroff chart**
	5.4′ × 2.7′	9.9	13	C31/D36

This galaxy has a mass of at least 70 billion Suns. It is part of the M 96 (NGC 3368) galaxy group – see M 96 (NGC 3368). Professional telescopic studies show an intergalactic cloud of atomic hydrogen (650 thousand light years in diameter) about 20′ SW of this galaxy that is thought to be the remains of a collision between this galaxy and M 96 (NGC 3368) that occurred 5 hundred million years ago. Although NGC 3384 forms an optical triplet with M 105 (8′ WSW – see M 105/NGC 3379) and NGC 3389 (spiral galaxy, 7′ SSE, mag. 11.9), NGC 3389 is actually a background object (about twice as far away) so that only M 105 and NGC 3384 are actual companions.

NGC 3395

Constellation	Object type	RA, Dec	Approx. transit date at local midnight	Distance
Leo Minor	Barred spiral galaxy	10h 49.8 m, +32° 59′	March 14	70 million light years
Age	Apparent size	Magnitude	Sky Atlas 2000.0 chart	Herald–Bobroff chart
	1.7′ × 0.9′	12.1	6	C31

This galaxy is interacting with its nearby companion NGC 3396 (barred spiral galaxy, 1′ E, mag. 12.1). A tail of atomic hydrogen 10′ in apparent length (2 hundred thousand light years in actual length) extends off to the SW of the NGC 3395/3396 pair in professional telescopic studies and is thought to be gas that was stripped from NGC 3395 in a previous close encounter with NGC 3396. The two galaxies are thought to be heading for another close encounter and are within about 50 million years of their closest approach. The present interaction appears to be driving intense star formation in both galaxies. The bridge between the two galaxies (visible in large amateur telescopes) is thought to be due to gravitational interaction rather than actual collision of material. Neither galaxy harbors an active galactic nucleus (see NGC 3079 for explanation). Both NGC 3395 and NGC 3396 are part of the NGC 3396 group of six gravitationally bound galaxies that also includes nearby NGC 3381 (peculiar barred spiral, 1° 44′ N, mag. 11.7), NGC 3424 (barred spiral, 26′ E, mag. 12.4), NGC 3430 (barred spiral, 30′ E, mag. 11.6) and NGC 3442 (spiral, 1° 10′ NE, mag. 13.4). Although also nearby, NGC 3413 (lenticular galaxy, 23′ SE, mag. 12.1) is actually a foreground object (at about half the distance to NGC 3395) and not part of this galaxy group.

NGC 3412

Constellation	Object type	RA, Dec	Approx. transit date at local midnight	Distance
Leo	Barred lenticular galaxy	10h 50.9 m, +13° 25′	March 14	35 million light years
Age	Apparent size	Magnitude	Sky Atlas 2000.0 chart	Herald–Bobroff chart
	3.7′ × 2.2′	10.5	13	C31/D36

It is part of the M 96 (NGC 3368) galaxy group. The population of young stars in the central region of this galaxy may be the result of the bar having recently driven disk material into the central region.

NGC 3414

Constellation	Object type	RA, Dec	Approx. transit date at local midnight	Distance
Leo Minor	Barred lenticular galaxy	10h 51.3 m, +27° 58′	March 14	60 million light years
Age	Apparent size	Magnitude	Sky Atlas 2000.0 chart	Herald–Bobroff chart
	3.6′ × 3.1′	11.0	6	C31

It has a low-ionization nuclear emission region ("LINER" – see M 81/NGC 3031). Professional telescopic studies find it has an odd shape, described as X-shaped or box-shaped, in addition to having faint extensions (like Saturn viewed edge-on) which may indicate an outer ring structure (a common feature of barred lenticular galaxies like this one).

NGC 3432

Constellation	Object type	RA, Dec	Approx. transit date at local midnight	Distance
Leo Minor	Barred spiral galaxy	10h 52.5 m, +36° 37'	March 14	25 million light years
Age	**Apparent size**	**Magnitude**	**Sky Atlas 2000.0 chart**	**Herald–Bobroff chart**
	6.6' × 1.6'	11.2	6	C31

This is an edge-on dwarf galaxy (its rotation axis is inclined from our line-of-sight by 79° and the SW side is receding from us in its rotation about this axis). It is distorted due to interactions with a nearby dwarf galaxy (UGC 5983, 4' WSW, mag. 15.5) that it is probably accreting. Tidal disturbances from its interaction with UGC 5983 are expected to cause intense star formation (a starburst) in the next few tens of millions of years. It has an optical diameter of about 50 thousand light years and a total dynamical mass of about 75 billion Suns.

NGC 3486

Constellation	Object type	RA, Dec	Approx. transit date at local midnight	Distance
Leo Minor	Barred spiral galaxy	11h 00.4 m, +28° 59'	March 16	25 million light years
Age	**Apparent size**	**Magnitude**	**Sky Atlas 2000.0 chart**	**Herald–Bobroff chart**
	6.8' × 4.8'	10.5	6	C31

It rotates about an axis inclined at about 40° from our line-of-sight. Its major axis is aligned at an angle of 80° counterclockwise from north. It has an optical diameter of about 50 thousand light years and a mass of about 30 billion Suns. It is a Seyfert galaxy (where a supermassive object in this galaxy's center accumulates nearby gas resulting in strong emission from the nucleus).

NGC 3489

Constellation	Object type	RA, Dec	Approx. transit date at local midnight	Distance
Leo	Barred lenticular galaxy	11h 00.3 m, +13° 54'	March 16	35 million light years
Age	**Apparent size**	**Magnitude**	**Sky Atlas 2000.0 chart**	**Herald–Bobroff chart**
	3.6' × 2.2'	10.3	13	C31/D36

It is part of the M 96 (NGC 3368) galaxy group (see M 96/NGC 3368).

NGC 3504

Constellation	Object type	RA, Dec	Approx. transit date at local midnight	Distance
Leo Minor	Barred spiral galaxy	11h 03.2 m, +27° 58'	March 17	65 million light years
Age	**Apparent size**	**Magnitude**	**Sky Atlas 2000.0 chart**	**Herald– Bobroff chart**
	2.7' × 2.5'	11.0	6	C31

Professional telescopic studies find intense star formation (a "starburst") is occurring in the center of this galaxy, which is thought to have begun about one hundred million years ago and may be occurring in a circumnuclear ring (of radius 1") associated with the inner of two so-called "inner Lindblad resonances" (where the speed of density waves resonantly amplifies oscillations in the orbits of matter in the disk). This galaxy is one of fewer than about a hundred galaxies that has had two or more recorded supernovae.

NGC 3521

Constellation	Object type	RA, Dec	Approx. transit date at local midnight	Distance
Leo	Barred spiral galaxy	11h 05.8 m, −00° 02'	March 18	25 million light years
Age	**Apparent size**	**Magnitude**	**Sky Atlas 2000.0 chart**	**Herald– Bobroff chart**
	11.2' × 5.4'	9.0	13	C49

This is a so-called "flocculent" spiral galaxy, meaning that it lacks any obvious azimuthally symmetric spiral arm pattern. This is generally attributed to the absence of strong spiral density waves. These waves normally travel around the disks of spiral galaxies, "firing up" stars as they go along giving rise to the spiral arms that are so obvious in "grand-design" spirals like M 51 (the process is mildly analogous to "the wave" at a large stadium if people took turns turning on flashlights instead of standing as "the wave" goes by). In NGC 3521, these waves are much weaker than usual and so only a weak spiral structure is found in professional telescopic studies. It has an optical diameter of about 80 thousand light years and a mass of about 150 billion Suns.

NGC 3556 (M 108)

Constellation	Object type	RA, Dec	Approx. transit date at local midnight	Distance
Ursa Major	Barred spiral galaxy	11h 11.5 m, +55° 40'	March 19	40 million light years
Age	**Apparent size**	**Magnitude**	**Sky Atlas 2000.0 chart**	**Herald–Bobroff chart**
	8.6' × 2.4'	10.0	2	C13/D35

Professional telescopic studies find that this galaxy has two giant loops of atomic hydrogen gas, one at the east and one at the west end. These loops have diameters of 10–20 thousand light years, have masses around 50 million Suns and are expanding outward but parallel to the disk of the galaxy at 40–50 km/s. They are thought to have originated about 50 million years ago. The loops may be the result of the rapid expansion of jets shot outward from an active galactic nucleus (powered by accretion onto a supermassive central black hole), with the jets "flaring" into shells as they reach the less dense outer regions of the galaxy. The nuclear activity has since largely subsided (since such activity is thought to last only some tens of millions of years in spirals). The galaxy is nearly edge-on (having an inclination angle of about 75°, meaning that it rotates about an axis inclined at 75° from our line-of-sight) and has an optical diameter of about 1 hundred thousand light years.

NGC 3587 (M 97)

Constellation	Object type	RA, Dec	Approx. transit date at local midnight	Distance
Ursa Major	Planetary nebula	11h 14.8 m, +55° 01'	March 20	2 thousand light years
Age	**Apparent size**	**Magnitude**	**Sky Atlas 2000.0 chart**	**Herald–Bobroff chart**
	3.2'	11	2	C12/D35

Nicknamed the "Owl Nebula" because it resembles two owl's eyes in a round disk when viewed at high power in large amateur telescopes. The nebula is expanding at a few tens of km/s and has a mass about 13% that of our Sun (not including the central star, whose mass is about 60% that of the Sun and is a challenging object in amateur telescopes). The three-dimensional structure of this nebula is complex in professional telescopic studies, although the "owl's eyes" are thought to be part of a bipolar ("hourglass") barrel-shaped structure (inclined from our line-of-sight by about 45°) that is inside three separate elliptical shells. The inner shell of the three gives rise to the round outer shape visible in amateur telescopes.

NGC 3593

Constellation	Object type	RA, Dec	Approx. transit date at local midnight	Distance
Leo	Spiral galaxy	11h 14.6 m, +12° 49′	March 20	40 million light years
Age	**Apparent size**	**Magnitude**	**Sky Atlas 2000.0 chart**	**Herald–Bobroff chart**
	4.9′ × 2.1′	10.9	13	C31/D36

Professional telescopic studies find that this galaxy possesses two stellar disks that are counter-rotating. This unusual situation may be the result of the accretion of a gas-rich dwarf satellite galaxy about a billion years ago which formed a counter-rotating circum-nuclear disk. The gas in this disk is fueling a nuclear starburst (i.e. intense star formation in the nucleus). At the current rate that this gas is formed into stars (about a star every year and a half), this activity will subside within about 6 hundred million years. It has an optical diameter of about 60 thousand light years and is part of the M 66 galaxy group (see M 66/NGC 3627).

NGC 3607

Constellation	Object type	RA, Dec	Approx. transit date at local midnight	Distance
Leo	Lenticular galaxy	11h 16.9 m, +18° 03′	March 21	70 million light years
Age	**Apparent size**	**Magnitude**	**Sky Atlas 2000.0 chart**	**Herald–Bobroff chart**
	4.6′ × 4.0′	9.9	13	C31/D36

This galaxy is thought to have acquired material from its neighbor NGC 3608 (6′ N – see NGC 3608) that has formed into a dust ring in the inner 10–20″ of NGC 3607. It is the namesake member of the NGC 3607 galaxy group whose membership is somewhat uncertain but is thought to include the following NGC galaxies among at least 15 galaxies: 3507 (barred spiral galaxy, 3° 11′ W, mag. 10.9), 3592 (spiral galaxy, 59′ SW, mag. 13.7), 3599 (lenticular galaxy, 22′ W, mag. 11.9), 3605 (elliptical galaxy, 2.5′ SW, mag. 12.3), 3608 (6′ NNE – see NGC 3608), 3626 (49′ ENE – see NGC 3626), 3655 (2° SE – see NGC 3655), 3659 (barred spiral galaxy, 1° 39′ E, mag. 12.3), 3681 (barred spiral, 2° 35′ ESE, mag. 11.2), 3684 (spiral, 2° 39′ ESE, mag. 11.4), 3686 (barred spiral, 2° 42′ ESE, mag. 11.3), 3691 (peculiar barred spiral, 2° 55′ ESE, mag. 11.8).

NGC 3608

Constellation	Object type	RA, Dec	Approx. transit date at local midnight	Distance
Leo	Elliptical galaxy	11h 17.0 m, +18° 09'	March 21	70 million light years
Age	**Apparent size**	**Magnitude**	**Sky Atlas 2000.0 chart**	**Herald–Bobroff chart**
	3.5' × 3.0'	10.8	13	C31/D36

It is part of the NGC 3607 galaxy group (see NGC 3607). The core of this galaxy counter-rotates from the rest of the galaxy, perhaps due to a previous merger.

NGC 3610

Constellation	Object type	RA, Dec	Approx. transit date at local midnight	Distance
Ursa Major	Elliptical galaxy	11h 18.4 m, +58° 47'	March 21	1 hundred million light years
Age	**Apparent size**	**Magnitude**	**Sky Atlas 2000.0 chart**	**Herald–Bobroff chart**
	2.5' × 2.5'	10.8	2	C13/D35

This galaxy is thought to have undergone a merger in the past few billion years, although whether the merger was of two similar sized disk galaxies (a "major merger") or merely the accretion of a minor satellite galaxy, remains an open question. Professional telescopes find an unusual "twisted disk" is present in the region within 2.7" of the center of this galaxy ("twisted" meaning the plane of the disk changes with radius). This galaxy is part of the NGC 3642 group of gravitationally bound galaxies that consists of NGC 3642 (spiral galaxy, 35' ENE, mag. 11.2), NGC 3619 (1° S – see NGC 3619), NGC 3674 (lenticular galaxy, 2° SSE, mag. 12.2), and NGC 3683 (barred spiral galaxy, 2° 15' SSE, mag. 12.4).

NGC 3613

Constellation	Object type	RA, Dec	Approx. transit date at local midnight	Distance
Ursa Major	Elliptical galaxy	11h 18.6 m, +58° 00'	March 21	1 hundred million light years
Age	**Apparent size**	**Magnitude**	**Sky Atlas 2000.0 chart**	**Herald–Bobroff chart**
	3.6' × 2.0'	10.9	2	C13/D35

It is the namesake member of the NGC 3613 group of four gravitationally bound galaxies that also includes NGC 3625 (barred spiral, 20' SE, mag. 13.1) and NGC 3669 (barred spiral, 57' ESE, mag. 12.4). Although NGC 3619 is nearby from our line-of-sight (15' SSE – see NGC 3619), it instead belongs to the NGC 3642 galaxy group (see NGC 3610).

NGC 3619

Constellation	Object type	RA, Dec	Approx. transit date at local midnight	Distance
Ursa Major	Lenticular galaxy	11h 19.4 m, +57° 45'	March 21	1 hundred million light years
Age	**Apparent size**	**Magnitude**	**Sky Atlas 2000.0 chart**	**Herald–Bobroff chart**
	3.5' × 2.4'	11.5	2	C13/D35

It is part of the NGC 3642 galaxy group (see NGC 3610). It has a total dynamical mass of about 70 billion Suns and contains an unusually large amount of hydrogen gas (about a billion Suns worth). This is unusual for a lenticular galaxy (which tend to be gas-poor) and may be due to a previous merger with a gas-rich galaxy. Although NGC 3613 is nearby from our line-of-sight (15' NNW – see NGC 3613) the two are not thought to be a bound pair since NGC 3613 instead belongs to its own separate galaxy group.

NGC 3621

Constellation	Object type	RA, Dec	Approx. transit date at local midnight	Distance
Hydra	Barred spiral galaxy	11h 18.3 m, −32° 49'	March 21	20 million light years
Age	**Apparent size**	**Magnitude**	**Sky Atlas 2000.0 chart**	**Herald–Bobroff chart**
	12.4' × 5.7'	9.7	20	C66

Present in this galaxy are many ionized hydrogen regions (HII regions), like M 42 (the Orion Nebula) or the Rosette Nebula (see NGC 2244). These regions are ionized by young stars (spectral type O and B) that formed within them. The galaxy has an optical diameter of about 70 thousand light years and an inclination angle of about 60° (90° is edge-on while 0° is face-on) and a position angle of 159° (i.e. its major axis is aligned SE–NW).

NGC 3623 (M 65)

Constellation	Object type	RA, Dec	Approx. transit date at local midnight	Distance
Leo	Barred spiral galaxy	11h 18.9 m, +13° 05'	March 21	35 million light years
Age	**Apparent size**	**Magnitude**	**Sky Atlas 2000.0 chart**	**Herald–Bobroff chart**
	9.0' × 2.3'	9.3	13	C49/D36

M 65 is nearly edge-on with its rotation axis inclined by 74° from our line-of-sight. It has an optical diameter of about 90 thousand light years. It is part of a gravitationally bound group of galaxies that includes NGC 3593 (1° SW), as well as nearby M 66 (see M 66/NGC 3627) and NGC 3628 (see NGC 3628) with which it forms the "Leo Triplet". Unlike M 66 and NGC 3628 (which have strongly interacted in the past – see NGC 3628), M 65 does not seem too much affected by the presence of the other galaxies in this group. M 65 is a "low-ionization nuclear emission region" (LINER) galaxy (see M 81/NGC 3031 for explanation).

NGC 3626

Constellation	Object type	RA, Dec	Approx. transit date at local midnight	Distance
Leo	Lenticular galaxy	11h 20.1 m, +18° 21'	March 21	70 million light years
Age	**Apparent size**	**Magnitude**	**Sky Atlas 2000.0 chart**	**Herald–Bobroff chart**
	3.2' × 2.4'	11.0	13	C31/D36

Gas in the central regions of this galaxy is counter-rotating relative to the stars, and is thought to be the result of a past minor merger with a gas-rich satellite galaxy. Although the optical diameter of the galaxy is about 65 thousand light years, professional telescopic studies find a ring of neutral hydrogen gas that extends out several times this distance. This gas may have been stripped from a gas-rich satellite galaxy as it was being accreted. NGC 3626 is part of the NGC 3607 galaxy group (see NGC 3607).

NGC 3627 (M 66)

Constellation	Object type	RA, Dec	Approx. transit date at local midnight	Distance
Leo	Barred spiral galaxy	11h 20.3 m, +12° 59'	March 21	35 million light years
Age	**Apparent size**	**Magnitude**	**Sky Atlas 2000.0 chart**	**Herald–Bobroff chart**
	9.1' × 4.1'	8.9	13	C49/D36

M 66 is part of the "Leo Triplet" that includes nearby M 65 (see M 65/NGC 3623) and NGC 3628, which form a gravitationally bound group of four galaxies (the fourth member being nearby NGC 3593). Professional telescopes find a quarter million light year (40') long plume of stars and gas extending to the east of NGC 3628, with a mass of hundreds of millions of Suns, that is thought to be the result of an interaction with M 66 nearly a billion years ago. M 66 is one of fewer than 20 galaxies that have had three or more recorded supernovae. It has an optical diameter of about 90 thousand light years.

NGC 3628

Constellation	Object type	RA, Dec	Approx. transit date at local midnight	Distance
Leo	Spiral galaxy	11h 20.3 m, +13° 35'	March 21	35 million light years
Age	**Apparent size**	**Magnitude**	**Sky Atlas 2000.0 chart**	**Herald–Bobroff chart**
	13.1' × 3.1'	9.5	13	C31/D36

This galaxy is nearly edge-on with an optical diameter of about 130 thousand light years. It forms a triplet with M 65 (NGC 3623) and M 66 (NGC 3627), with which it forms a gravitationally bound group of galaxies that also includes NGC 3593 (1° 36' WSW – see NGC 3593). Intense star formation (a "starburst") is occurring in the nucleus of NGC 3628 which blows material out of the center in a collimated outflow perpendicular to the plane of the galaxy (see M 82 for further explanation). An X-ray point source is offset 20" WSW (several thousand light years) from the nucleus of the galaxy. The origin of this source remains unexplained, but may be due to accretion onto an intermediate sized black hole (with a mass of a thousand or so Suns) that is separate from the galaxy nucleus. Professional telescopes also find a quarter million light year (40') long plume of stars and gas extending to the east of NGC 3628 that is thought to be the result of an interaction with M 66 nearly a billion years ago.

NGC 3631

Constellation	Object type	RA, Dec	Approx. transit date at local midnight	Distance
Ursa Major	Spiral galaxy	11h 21.0 m, +53° 10′	March 22	60 million light years
Age	**Apparent size**	**Magnitude**	**Sky Atlas 2000.0 chart**	**Herald–Bobroff chart**
	5.0′ × 4.8′	10.4	2	C12/D35

This is a "grand-design" spiral, meaning that it has two symmetrically placed spiral arms that extend over most of its visible disk in professional telescopic images. The radius of corotation (where the waves responsible for the spiral arms are traveling at the same speed as the disk matter itself) is situated 42″ from the galaxy center. The galaxy is nearly face-on (inclination angle of 17° – exactly face-on would be 0°). It is one of fewer than about 20 galaxies that have had at least three recorded supernovae. It is the namesake member of the NGC 3631 group of eight gravitationally bound galaxies, including NGC 3718 (peculiar barred spiral galaxy, 1° 45′ E, mag. 10.8), NGC 3729 (1° 55′ E – see NGC 3729), NGC 3913 (spiral, 4° 50′ ENE, mag. 12.6), NGC 3972 (barred spiral, 5° 31′ ENE, mag. 12.3) and NGC 3998 (5° 50′ ENE – see NGC 3998).

NGC 3640

Constellation	Object type	RA, Dec	Approx. transit date at local midnight	Distance
Leo	Elliptical galaxy	11h 21.1 m, +03° 14′	March 22	80 million light years
Age	**Apparent size**	**Magnitude**	**Sky Atlas 2000.0 chart**	**Herald–Bobroff chart**
	4.5′ × 4.0′	10.4	13	C49

It is the namesake member of the NGC 3640 galaxy group that contains six gravitationally bound galaxies, including NGC 3630 (lenticular galaxy, 20′ SW, mag. 11.9), NGC 3664 (peculiar barred spiral galaxy, 50′ E, mag. 12.8), NGC 3611 (spiral galaxy, 1° 36′ NW, mag. 12.1) and NGC 3641 (elliptical galaxy, 2.5′ SE, mag. 13.2). For an elliptical galaxy, NGC 3640 rotates at an unusually high velocity (up to 120 km/s), about its major axis (which is aligned E–W). This galaxy is also interacting with NGC 3641. This interaction may be causing patchy shell-like structures in NGC 3640, but is not thought to be responsible for the fast rotation of NGC 3640. Instead, this rotation may be the result of a past merger (and accretion of) a disk galaxy.

NGC 3655

Constellation	Object type	RA, Dec	Approx. transit date at local midnight	Distance
Leo	Spiral galaxy	11h 22.9 m, +16° 35′	March 22	70 million light years
Age	Apparent size	Magnitude	Sky Atlas 2000.0 chart	Herald–Bobroff chart
	1.5′ × 1.0′	11.6	13	C31/D36

It is part of the NGC 3607 galaxy group (see NGC 3607). It has an optical diameter of about 30 thousand light years, an inclination angle of about 50° (meaning its axis of rotation is inclined at 50° from our line-of-sight) and a position angle of 30° (meaning its major axis is aligned 30° counterclockwise from N).

NGC 3665

Constellation	Object type	RA, Dec	Approx. transit date at local midnight	Distance
Ursa Major	Lenticular galaxy	11h 24.7 m, +38° 46′	March 23	90 million light years
Age	Apparent size	Magnitude	Sky Atlas 2000.0 chart	Herald–Bobroff chart
	3.5′ × 3.0′	10.8	6	C31

This galaxy is thought to be oblate in shape (meaning it is shaped like a squashed tennis ball). It has a mass of about 3 hundred billion Suns. A thin dust lane is present in professional telescopic studies. Two symmetric radio jets (due to synchrotron emission of high-speed electrons gyrating wildly in a magnetic field) emanate perpendicular to this dust lane near the center of the galaxy, extending outward for 1.5′ (40 thousand light years) and thought to be confined by a gaseous halo. The jets are thought to be driven by an active galactic nucleus (in which energetic emission is caused by accretion onto a massive central object e.g. a black hole). It is the namesake member of the NGC 3665 group of gravitationally bound galaxies that includes five galaxies, among them being NGC 3648 (lenticular galaxy, 1° 11′ NNW, mag. 12.6), NGC 3652 (peculiar barred spiral, 1° 5′ SSW, mag. 12.2) and NGC 3658 (lenticular galaxy, 15′ SW, mag. 12.2).

NGC 3675

Constellation	Object type	RA, Dec	Approx. transit date at local midnight	Distance
Ursa Major	Spiral galaxy	11h 26.1 m, +43° 35′	March 23	40 million light years
Age	Apparent size	Magnitude	Sky Atlas 2000.0 chart	Herald–Bobroff chart
	6.2′ × 3.6′	10.2	6	C30/D35

This galaxy has a low-ionization nuclear emission region ("LINER" – see M 81/NGC 3031 for explanation). It has an optical diameter of about 70 thousand light years, a mass of 2 hundred billion Suns and an inclination angle of about 60°.

NGC 3686

Constellation	Object type	RA, Dec	Approx. transit date at local midnight	Distance
Leo	Barred spiral galaxy	11h 27.7 m, +17° 13′	March 23	70 million light years
Age	**Apparent size**	**Magnitude**	**Sky Atlas 2000.0 chart**	**Herald–Bobroff chart**
	3.1′ × 2.4′	11.3	13	C30/D36

It forms a subgroup of four galaxies with NGC 3681, NGC 3684, NGC 3691 that are all part of the NGC 3607 group of galaxies (see NGC 3607). These four galaxies move about in their subgroup, with an average time for one of them to traverse the space enclosed by their subgroup (called the "crossing time") being about a billion years. NGC 3686 has an optical diameter of about 60 thousand light years.

NGC 3726

Constellation	Object type	RA, Dec	Approx. transit date at local midnight	Distance
Ursa Major	Barred spiral galaxy	11h 33.3 m, +47° 02′	March 25	50 million light years
Age	**Apparent size**	**Magnitude**	**Sky Atlas 2000.0 chart**	**Herald–Bobroff chart**
	6.0′ × 4.1′	10.4	6	C12/D35

It has a mass of one or 2 hundred billion Suns and an optical diameter of about 90 thousand light years. Its inclination angle is about 45° (i.e. it rotates about an axis tilted at 45° to our line-of-sight) and its position angle is 10° (i.e. its major axis is aligned 10° counterclockwise from N–S). It is part of the Ursa Major galaxy cluster (see M 109/NGC 3992) as well as the NGC 3992 (M 109) galaxy group.

NGC 3729

Constellation	Object type	RA, Dec	Approx. transit date at local midnight	Distance
Ursa Major	Peculiar barred spiral galaxy	11h 33.8 m, +53° 08′	March 25	60 million light years
Age	**Apparent size**	**Magnitude**	**Sky Atlas 2000.0 chart**	**Herald–Bobroff chart**
	2.9′ × 1.9′	11.4	2	C12/D35

It has a total mass of about 40 billion Suns and an optical diameter of about 50 thousand light years. It is part of the NGC 3631 galaxy group (see NGC 3631), as is its close neighbor NGC 3718 (12′ WSW), the latter having a "warped disk" (meaning the central parts of its disk lie in a different plane than the outer regions). NGC 3718 has about 10 times the total mass of NGC 3729 (giving it a total mass of about 4 hundred billion Suns). NGC 3729 and NGC 3718 are thought to be an interacting pair.

NGC 3810

Constellation	Object type	RA, Dec	Approx. transit date at local midnight	Distance
Leo	Spiral galaxy	11h 41.0 m, +11° 28'	March 27	40 million light years
Age	Apparent size	Magnitude	Sky Atlas 2000.0 chart	Herald–Bobroff chart
	4.1' × 2.7'	10.8	13	C48

It has a mass of about 50 billion Suns and an optical diameter of about 50 thousand light years. NGC 3773 (lenticular galaxy, 57' NW, mag. 12.0) is thought to be a companion to this galaxy.

NGC 3813

Constellation	Object type	RA, Dec	Approx. transit date at local midnight	Distance
Ursa Major	Spiral galaxy	11h 41.3 m, +36° 33'	March 27	90 million light years
Age	Apparent size	Magnitude	Sky Atlas 2000.0 chart	Herald–Bobroff chart
	2.0' × 1.0'	11.6	6	C30

It has an optical diameter of about 50 thousand light years, a mass of about 40 billion Suns, an inclination angle of about 60° and a position angle of 87° (meaning its major axis is aligned E–W).

NGC 3877

Constellation	Object type	RA, Dec	Approx. transit date at local midnight	Distance
Ursa Major	Spiral galaxy	11h 46.1 m, +47° 30'	March 28	50 million light years
Age	Apparent size	Magnitude	Sky Atlas 2000.0 chart	Herald–Bobroff chart
	5.3' × 1.2'	11.0	6	C12/D35

A foreground star (in our Galaxy) superimposed on the nucleus gives the center of this galaxy a stellar appearance in large amateur telescopes. The galaxy is nearly edge-on (with an inclination angle of 84°) and has an optical diameter of about 80 thousand light years. It is part of the Ursa Major galaxy cluster (see M 109/NGC 3992) as well as the NGC 3992 galaxy group.

NGC 3893

Constellation	Object type	RA, Dec	Approx. transit date at local midnight	Distance
Ursa Major	Barred spiral galaxy	11h 48.6 m, +48° 43′	March 29	50 million light years
Age	**Apparent size**	**Magnitude**	**Sky Atlas 2000.0 chart**	**Herald–Bobroff chart**
	4.5′ × 2.4′	10.5	6	C12/D35

This is a "grand-design" spiral, meaning that it has two symmetrically placed spiral arms that extend over most its visible disk in professional telescopic images. The inner 50% (in area) of this galaxy contains a mass of about 30 billion Suns. It forms a pair with NGC 3896 (barred lenticular galaxy, 4′ SE, mag. 12.9) whose close presence is thought to have caused a higher rotation velocity in the north of NGC 3893 than in the south. It is part of the Ursa Major galaxy cluster (see M 109/NGC 3992), as well as the NGC 3992 galaxy group.

NGC 3898

Constellation	Object type	RA, Dec	Approx. transit date at local midnight	Distance
Ursa Major	Spiral galaxy	11h 49.3 m, +56° 05′	March 29	70 million light years
Age	**Apparent size**	**Magnitude**	**Sky Atlas 2000.0 chart**	**Herald–Bobroff chart**
	3.8′ × 2.6′	10.7	2	C12/D35

This is a "flocculent" spiral galaxy (see NGC 3521). With several billion solar masses worth of neutral hydrogen gas, this galaxy is quite gas-rich for its galaxy type (it is an "early-type" spiral), although the density of this gas is well below that needed for spontaneous star formation from gravitational collapse. Thus, other, local mechanisms are thought to be responsible for the formation of the young, bright stars seen in its flocculent pattern, although without the help of the gravitational collapse of gas clouds, star formation rates in this galaxy are much lower than may have occurred when more gas was perhaps present earlier in this galaxy's life. The total mass of the galaxy is several hundred billion Suns and its optical diameter is about 80 thousand light years. It is the namesake member of the NGC 3898 group of seven gravitationally bound galaxies that includes NGC 3733 (barred spiral, 2° 22′ WSW, mag. 12.4), NGC 3756 (barred spiral, 2° 31′ SW, mag. 11.5), NGC 3804 (barred spiral, 1° 10′ W, mag. 12.9), NGC 3850 (barred spiral, 33′ WSW, mag. 13.3), NGC 3982 (barred spiral, 1° 24′ SE, mag. 11.0), and NGC 3846A (irregular barred, 1° 16′ SW, mag. 13.3, labeled as 3846 on Herald–Bobroff chart C12 and unlabelled on chart D35).

NGC 3900

Constellation	Object type	RA, Dec	Approx. transit date at local midnight	Distance
Leo	Lenticular galaxy	11h 49.2 m, +27° 01′	March 29	90 million light years
Age	**Apparent size**	**Magnitude**	**Sky Atlas 2000.0 chart**	**Herald–Bobroff chart**
	3.5′ × 1.8′	11.4	6	C30

It has an optical diameter of about 90 thousand light years. It is thought to be part of a gravitationally bound group of three galaxies that includes NGC 3912 (34′ SSE – see NGC 3912) and the much dimmer UGC 6791 (spiral, 17′ S, mag. 14). NGC 3900 is believed to have undergone a minor merger in its past, the remains of which are seen in the unusual motions of some of the gas in this galaxy in professional telescopic studies.

NGC 3912

Constellation	Object type	RA, Dec	Approx. transit date at local midnight	Distance
Leo	Barred spiral galaxy	11h 50.1 m, +26° 29′	March 29	90 million light years
Age	**Apparent size**	**Magnitude**	**Sky Atlas 2000.0 chart**	**Herald–Bobroff chart**
	1.5′ × 0.9′	12.4	6	C30

It is part of the NGC 3900 galaxy group (see NGC 3900). It has an optical diameter of about 40 thousand light years, an inclination angle of about 60°, and a position angle of 5°. NGC 3899 is a duplicate entry of NGC 3912.

NGC 3938

Constellation	Object type	RA, Dec	Approx. transit date at local midnight	Distance
Ursa Major	Spiral galaxy	11h 52.6 m, +44° 07′	March 30	60 million light years
Age	**Apparent size**	**Magnitude**	**Sky Atlas 2000.0 chart**	**Herald–Bobroff chart**
	5.1′ × 5.0′	10.4	6	C12/D35

It is nearly face-on (with an inclination angle of 14°) and has an optical diameter of about 90 thousand light years. Most of the disk rotates at about 150 km/s. This galaxy is part of the Ursa Major galaxy cluster (see M 109/NGC 3992) and the NGC 4051 subgroup of this cluster (see NGC 4051).

NGC 3941

Constellation	Object type	RA, Dec	Approx. transit date at local midnight	Distance
Ursa Major	Barred lenticular galaxy	11h 52.8 m, +36° 59'	March 30	50 million light years
Age	Apparent size	Magnitude	Sky Atlas 2000.0 chart	Herald–Bobroff chart
	3.5' × 2.5'	10.3	6	C30

An extended disk of counter-rotating gas found in professional telescopic studies suggests that this galaxy underwent a merger or accretion with another galaxy in its past. It is a Seyfert galaxy (where a supermassive object in this galaxy's center accumulates nearby material and produces strong emission).

NGC 3945

Constellation	Object type	RA, Dec	Approx. transit date at local midnight	Distance
Ursa Major	Barred lenticular galaxy	11h 53.2 m, +60° 40'	March 30	70 million light years
Age	Apparent size	Magnitude	Sky Atlas 2000.0 chart	Herald–Bobroff chart
	5.0' × 3.0'	10.9	2	C12/D35

Professional telescopic studies find two nested bars in this galaxy (one about 5" in length, the other about 80" in length and surrounded by inner and outer rings, the latter being common in barred lenticular galaxies – see NGC 2859). The bars are at different position angles and probably rotate independently of each other. Bars in galaxies can transport material from the disk into the nucleus, thereby fueling activity in the nucleus. Whether this is what fuels the LINER (low-ionization nuclear emission region – see M 81/NGC 3031 for explanation) in the nucleus of this galaxy is not known.

NGC 3949

Constellation	Object type	RA, Dec	Approx. transit date at local midnight	Distance
Ursa Major	Spiral galaxy	11h 53.7 m, +47° 52'	March 30	60 million light years
Age	Apparent size	Magnitude	Sky Atlas 2000.0 chart	Herald–Bobroff chart
	2.9' × 1.7'	11.1	6	C12/D35

It is part of the Ursa Major galaxy cluster (see M 109/NGC 3992) as well as the NGC 3992 galaxy group. It has a mass of about 30 billion Suns, an optical diameter of about 50 thousand light years, an inclination angle of about 55°, and a position angle of 120° (SE–NW).

NGC 3953

Constellation	Object type	RA, Dec	Approx. transit date at local midnight	Distance
Ursa Major	Barred spiral galaxy	11h 53.8 m, +52° 19′	March 30	60 million light years
Age	**Apparent size**	**Magnitude**	**Sky Atlas 2000.0 chart**	**Herald–Bobroff chart**
	6.9′ × 3.6′	10.1	2	C12/D35

It is part of the Ursa Major galaxy cluster (see M 109/NGC 3992) as well as the NGC 3992 galaxy group (see M 109/NGC 3992). It has a mass of about 140 billion Suns, an optical diameter of about 120 thousand light years, an inclination angle of about 60° and a position angle of 13°. It is a LINER (low-ionization nuclear emission region) galaxy (see M 81/NGC 3031) with ionized hydrogen (HII) regions (like M 42).

NGC 3962

Constellation	Object type	RA, Dec	Approx. transit date at local midnight	Distance
Crater	Elliptical galaxy	11h 54.7 m, −13° 59′	March 31	80 million light years
Age	**Apparent size**	**Magnitude**	**Sky Atlas 2000.0 chart**	**Herald–Bobroff chart**
	3.4 ′ × 2.8′	10.7	13	C48

This galaxy contains an inner disk of ionized gas with several tens of thousands of solar masses, which is in contrast to the original notion that elliptical and lenticular (so-called "early-type") galaxies contain no interstellar matter. However, it is now recognized that about half of early-type galaxies contain significant amounts of dust and gas, one suggested source of this material being accretion from past mergers with other galaxies. Indeed, professional telescopic studies find an odd arc-like structure of ionized gas in NGC 3962 that may have been accreted in an earlier merger with another galaxy.

NGC 3982

Constellation	Object type	RA, Dec	Approx. transit date at local midnight	Distance
Ursa Major	Barred spiral galaxy	11h 56.5 m, +55° 07′	March 31	70 million light years
Age	**Apparent size**	**Magnitude**	**Sky Atlas 2000.0 chart**	**Herald–Bobroff chart**
	2.3′ × 2.0′	11.0	2	C12/D35

It is nearly face-on (with an inclination angle of 26°). It is part of the NGC 3898 galaxy group (see NGC 3898), as well as the Ursa Major galaxy cluster (see M 109/NGC 3992). This galaxy is a Seyfert galaxy (see NGC 3227) and has a circumnuclear ring of very active star formation (at radii 5–20″ from the center, with many ionized hydrogen HII regions). This is also a multiple-armed spiral (as opposed to a "grand-design" spiral which has only two symmetrically placed arms, or a "flocculent" spiral where star formation occurs in patches throughout the disk with no obvious spiral arms).

NGC 3992 (M 109)

Constellation	Object type	RA, Dec	Approx. transit date at local midnight	Distance
Ursa Major	Barred spiral galaxy	11h 57.6 m, +53° 22'	March 31	60 million light years
Age	**Apparent size**	**Magnitude**	**Sky Atlas 2000.0 chart**	**Herald–Bobroff chart**
	7.5' × 4.4'	9.8	2	C12/D35

It has a mass of about 250 billion Suns and a diameter of about 130 thousand light years. It is part of the Ursa Major galaxy cluster, one of only three major galaxy clusters within 150 million light years of us (the others being the Virgo cluster and the Fornax cluster). Galaxy clusters are the largest stable structures in the Universe. Only about 1% of galaxies belong to galaxy clusters. The Ursa Major cluster contains about 80 known galaxies, and has about 5% of the mass (but emits 30% of the light) of the Virgo cluster (see M 49/NGC 4472). The Ursa Major cluster is an unusual cluster in that it consists almost entirely of late-type galaxies (e.g. Sc and SBc and later galaxies in Hubble's galaxy classification scheme; Sc and SBc spirals are "late-type" spirals that have prominent, loosely wound, spiral arms and only a very small central bulge relative to an extended disk). In contrast, most galaxy clusters consist of "early-type" galaxies (i.e. elliptical and lenticular, and early-type spirals, the latter having relatively tightly wound spiral arms and a large central bulge). Indeed, 3/4 of the Virgo cluster galaxies are early-type. M 109 is the namesake member of the M 109/NGC 3992 galaxy group of gravitationally bound galaxies that is a subgroup of the Ursa Major cluster and consists of 27 galaxies.

NGC 3998

Constellation	Object type	RA, Dec	Approx. transit date at local midnight	Distance
Ursa Major	Lenticular galaxy	11h 57.9 m, +55° 27'	March 31	60 million light years
Age	**Apparent size**	**Magnitude**	**Sky Atlas 2000.0 chart**	**Herald–Bobroff chart**
	2.7' × 2.3'	10.7	2	C12/D35

This galaxy is part of the NGC 3631 galaxy group (see NGC 3631), as well as the Ursa Major galaxy cluster (see M 109/NGC 3992). NGC 3998 is a LINER ("low-ionization nuclear emission region") galaxy (see M 81/NGC 3031), which although in general may be due to several possible mechanisms, in this case is thought to be one of at least about 20% of LINERs where the nuclear emission is due to accretion onto a supermassive black hole. The mass of the central black hole in this galaxy is thought to be about 6 hundred million Suns. The presence of some interstellar matter (dust and gas) in NGC 3998 is thought to be due to accretion of a dwarf galaxy about a billion years ago. NGC 3990 (lenticular galaxy, 5' W, mag. 12.6) is nearby both apparently and physically, but the two galaxies show only weak interaction.

NGC 4026

Constellation	Object type	RA, Dec	Approx. transit date at local midnight	Distance
Ursa Major	Lenticular galaxy	11h 59.4 m, +50° 58′	April 1	60 million light years
Age	**Apparent size**	**Magnitude**	**Sky Atlas 2000.0 chart**	**Herald–Bobroff chart**
	4.7′ × 1.2′	10.8	2	C12/D35

It is part of the Ursa Major galaxy cluster (see M 109/NGC 3922) and the NGC 3992 galaxy group (see M 109/NGC 3992). It is nearly edge-on (with an inclination angle of 80°) and has an optical diameter of about 80 thousand light years and a total mass of about 80 billion Suns. Professional telescopic studies find several filaments of neutral hydrogen (HI) gas that are each over a hundred thousand light years long (containing billions of solar masses), lying in the neighborhood of NGC 4026 and its nearby companions UGC 6917 (barred spiral, 42′ SW, mag. 12), and the much dimmer UGC 6922 and UGC 6956. This gas is expected to be captured by NGC 4026 over the coming millions to billions of years, possibly forming into a ring around this galaxy.

NGC 4027

Constellation	Object type	RA, Dec	Approx. transit date at local midnight	Distance
Corvus	Barred spiral galaxy	11h 59.5 m, −19° 16′	April 1	70 million light years
Age	**Apparent size**	**Magnitude**	**Sky Atlas 2000.0 chart**	**Herald–Bobroff chart**
	3.3′ × 2.4′	11.1	13	C66

This galaxy is an oddity in professional telescopic images because it is a one-armed spiral (its northern arm is far stronger than its southern arm) and its prominent bar is off-center. NGC 4027A (mag. 14) lies 4′ S. The two galaxies are interacting – indeed the dark matter halo of NGC 4027 extends out to NGC 4027A. Their interaction is thought to have caused the lopsidedness of NGC 4027's spiral arms, as well as a ring of atomic hydrogen (HI) gas around NGC 4027 (in professional telescopic studies) that has been stripped from NGC 4027A. NGC 4027A orbits NGC 4027 with a period of about 5 hundred million years and is about 250 million years from its last closest approach. NGC 4027 has a mass of about 60 billion Suns and is part of the NGC 4038 galaxy group (see NGC 4038).

NGC 4030

Constellation	Object type	RA, Dec	Approx. transit date at local midnight	Distance
Virgo	Spiral galaxy	12h 00.4 m, −01° 06′	April 15 (Daylight Savings Time)	80 million light years
Age	**Apparent size**	**Magnitude**	**Sky Atlas 2000.0 chart**	**Herald–Bobroff chart**
	4.2′ × 3.2′	10.6	13	C48/D34

It has an optical diameter of about 1 hundred thousand light years, a mass of nearly 2 hundred billion Suns, an inclination angle of about 45° and a position angle of 27°. It is the namesake member of the NGC 4030 group of four gravitationally bound galaxies, the other members being non-NGC galaxies, the brightest of which is UGC 7000 (irregular barred galaxy, mag. 14, 17′ SE).

NGC 4036

Constellation	Object type	RA, Dec	Approx. transit date at local midnight	Distance
Ursa Major	Lenticular galaxy	12h 01.5 m, +61° 54′	April 16	80 million light years
Age	**Apparent size**	**Magnitude**	**Sky Atlas 2000.0 chart**	**Herald–Bobroff chart**
	4.0′ × 1.8′	10.7	2	C12/D35

This is a LINER (low-ionization nuclear emission region) galaxy (see M 81/NGC 3031). It is the namesake member of the NGC 4036 group of three gravitationally bound galaxies that includes NGC 4041 (15′ NNE – see NGC 4041) and IC 758 (barred spiral, 41′ NNE, mag. 13). NGC 4036 is nearly edge-on (inclination angle 71°).

NGC 4038

Constellation	Object type	RA, Dec	Approx. transit date at local midnight	Distance
Corvus	Barred spiral galaxy	12h 01.9 m, −18° 52′	April 16	70 million light years
Age	**Apparent size**	**Magnitude**	**Sky Atlas 2000.0 chart**	**Herald–Bobroff chart**
	3.4′ × 2.0′	10.3	13	C66

Together with NGC 4039 this is nicknamed the "Ring-Tail Galaxy" or the "Antennae". The two disk galaxies are colliding and form an odd comma shape in the sky. The merger is causing intense star formation (perhaps five stars per year) including hundreds of compact young star clusters (which in many billions of years will be like globular clusters in our Galaxy). These star clusters occur throughout the two galaxies and have ages of a few million years (except in the nuclei where they are about 10 times older), with masses up to several million Suns. Professional telescopes find that debris ejected in the collision of these two galaxies has apparently birthed a dwarf galaxy 10′ to the SW. Several tens of X-ray point-sources are found in these two galaxies, a few of which are associated with starburst regions, but most are thought to be associated with black holes with masses of several tens of Suns. This galaxy is the namesake member of the NGC 4038 galaxy group that includes 20 nearby gravitationally bound galaxies.

NGC 4041

Constellation	Object type	RA, Dec	Approx. transit date at local midnight	Distance
Ursa Major	Spiral galaxy	12h 02.2 m, +62° 08′	April 16	80 million light years
Age	**Apparent size**	**Magnitude**	**Sky Atlas 2000.0 chart**	**Herald–Bobroff chart**
	2.7′ × 2.6′	11.3	2	C12/D35

This is a multiple-arm spiral (see NGC 3982 for explanation) and part of the NGC 4036 galaxy group (see NGC 4036). It is about 60 thousand light years in diameter, and nearly face-on (with an inclination angle of about 20°).

NGC 4051

Constellation	Object type	RA, Dec	Approx. transit date at local midnight	Distance
Ursa Major	Barred spiral galaxy	12h 03.2 m, +44° 32'	April 16	60 million light years
Age	**Apparent size**	**Magnitude**	**Sky Atlas 2000.0 chart**	**Herald–Bobroff chart**
	5.3' × 4.4'	10.2	7	C12/D35

The SW side is the nearer side of this galaxy. It is a Seyfert galaxy and one of twelve such galaxies so identified by Seyfert himself. Such galaxies have a supermassive object in the center (about 1.5 million solar masses for this galaxy) that accumulate nearby material, resulting in strong emission from the nucleus. Nearby NGC 4013 (spiral galaxy, 1° SW, mag. 11.2) has been suggested as a companion galaxy that tidally disturbed NGC 4051 in the past, creating gravitational instability in the disk of NGC 4051 that led to its Seyfert character. NGC 4051 is part of the Ursa Major galaxy cluster (see M 109/NGC 3992) and is the namesake member of the NGC 4051 group of 16 galaxies that includes nearby NGC 3906 (4° 32' NNW), NGC 3938 (1° 54' WSW – see NGC 3938), IC 750 (2° SSW), NGC 4143 (2° 19' SSE – see NGC 4143), NGC 4096 (3° N), NGC 4111 (1° 38' SSE – see NGC 4111), NGC 4117 (1° 38' SSE), NGC 4138 (1° 25' SE), NGC 4183 (2° ESE), NGC 4218 (4° 12', NNE), NGC 4389 (4° 07' ENE), NGC 4288 (3° 31' ENE) and NGC 4346 (4° 19' NE – see NGC 4346.)

NGC 4085

Constellation	Object type	RA, Dec	Approx. transit date at local midnight	Distance
Ursa Major	Barred spiral galaxy	12h 05.4 m, +50° 21'	April 17	60 million light years
Age	**Apparent size**	**Magnitude**	**Sky Atlas 2000.0 chart**	**Herald–Bobroff chart**
	2.5' × 0.8'	12.4	2	C12/D35

It is part of the Ursa Major galaxy cluster as well as the NGC 3992 galaxy group (see M 109/NGC 3992). It forms a pair with NGC 4088 (see NGC 4088, 12' N). It has an optical diameter of about 40 thousand light years.

NGC 4088

Constellation	Object type	RA, Dec	Approx. transit date at local midnight	Distance
Ursa Major	Barred spiral galaxy	12h 05.6 m, +50° 32'	April 17	60 million light years
Age	**Apparent size**	**Magnitude**	**Sky Atlas 2000.0 chart**	**Herald–Bobroff chart**
	5.6' × 2.1'	10.6	2	C12/D35

It is part of the Ursa Major galaxy cluster and the NGC 3992 galaxy group (see M 109/NGC 3992). It rotates about an axis inclined at about 70° from our line-of-sight, with the NE portion of the galaxy rotating much more slowly (at about one hundred km/s) than the SW portion of the galaxy (where the rotation speed is about 2 hundred km/s). One explanation for this asymmetry is tidal interaction with nearby NGC 4085 (12' S – see NGC 4085). NGC 4088 has an optical diameter of about 1 hundred thousand light years.

NGC 4102

Constellation	Object type	RA, Dec	Approx. transit date at local midnight	Distance
Ursa Major	Barred spiral galaxy	12h 06.4 m, +52° 43′	April 17	60 million light years
Age	**Apparent size**	**Magnitude**	**Sky Atlas 2000.0 chart**	**Herald–Bobroff chart**
	3.1′ × 1.7′	11.2	2	C12/D35

It is part of the Ursa Major galaxy cluster (see M 109/NGC 3992). This galaxy has intense star formation occurring in its nucleus (a "starburst") and is also a LINER (see M 81/NGC 3031) or weak Seyfert galaxy (see NGC 3227). It has an optical diameter of about 50 thousand light years.

NGC 4111

Constellation	Object type	RA, Dec	Approx. transit date at local midnight	Distance
Canes Venatici	Lenticular galaxy	12h 07.0 m, +43° 04′	April 17	60 million light years
Age	**Apparent size**	**Magnitude**	**Sky Atlas 2000.0 chart**	**Herald–Bobroff chart**
	4.6′ × 1.0′	10.7	7	C12/D35

It is part of the Ursa Major galaxy cluster (see M 108/NGC 3992) and the NGC 4051 galaxy group (see NGC 4051). This is a LINER galaxy (see M 81/NGC 3031) whose nuclear emission is thought to be due to the presence of young, hot stars in its center (rather than photoionization associated with accretion onto a central black hole as in M 81/NGC 3031).

NGC 4143

Constellation	Object type	RA, Dec	Approx. transit date at local midnight	Distance
Canes Venatici	Barred lenticular galaxy	12h 09.6 m, +42° 32′	April 18	60 million light years
Age	**Apparent size**	**Magnitude**	**Sky Atlas 2000.0 chart**	**Herald–Bobroff chart**
	2.9′ × 1.9′	10.6	7	C12/D35

It is part of the Ursa Major galaxy cluster and the NGC 4051 subgroup of this cluster (see NGC 4051). It is a LINER galaxy (see M 81/NGC 3031).

NGC 4147

Constellation	Object type	RA, Dec	Approx. transit date at local midnight	Distance
Coma Berenices	Globular cluster	12h 10.1 m, +18° 33′	April 18	60 thousand light years
Age	**Apparent size**	**Magnitude**	**Sky Atlas 2000.0 chart**	**Herald–Bobroff chart**
	4.1′	10.3	13	C30/D33

This globular cluster has a mass of about 70 thousand Suns and its size corresponds to a diameter of about 70 light years. Like almost all other globular clusters in our Galaxy, this cluster doesn't orbit with the material in the Galaxy's disk. Instead it spends about 2/3 of its time at distances farther than 130 thousand light years from the Galactic center (out in the halo of the Galaxy) in a precessing path that orbits the Galaxy about once every 1/2 billion years and is inclined to the Galactic central plane by about 45°.

NGC 4150

Constellation	Object type	RA, Dec	Approx. transit date at local midnight	Distance
Coma Berenices	Lenticular galaxy	12h 10.6 m, +30° 24′	April 18	15 million light years
Age	**Apparent size**	**Magnitude**	**Sky Atlas 2000.0 chart**	**Herald–Bobroff chart**
	2.3′ × 1.6′	11.6	7	C30

It probably belongs to the Canes Venatici galaxy cloud, a collection of about 50 spiral and irregular galaxies lying in Canes Venatici between +20° and +60° declination and 11.5h to 14h R.A. that contains about a quarter of the 215 known galaxies with radial velocities < 5 hundred km/s. Most of the galaxies in this cloud are about 10–30 million light years away.

NGC 4151

Constellation	Object type	RA, Dec	Approx. transit date at local midnight	Distance
Canes Venatici	Barred spiral galaxy	12h 10.5 m, +39° 24′	April 18	50 million light years
Age	**Apparent size**	**Magnitude**	**Sky Atlas 2000.0 chart**	**Herald–Bobroff chart**
	6.8′ × 5.3′	10.8	7	C30

It has an optical diameter of about 1 hundred thousand light years. This is a Seyfert galaxy, where nuclear emission is thought to be due to accretion onto a central black hole. The nuclear emission from this galaxy varies considerably on different time scales, the mechanisms for which are not certain, although one possible mechanism involves compressible flow instabilities, involving shocks in jets, in the relativistic mini-jets that are thought to beam out from the center of this galaxy. The micro-variations on shorter time scales (within one night) are thought to be associated with a "blazar" in which the jet is directed along our line-of-sight at relativistic speeds. Because of relativistic Doppler effects, variations in the jet appear compressed in time from our reference frame, making them appear to us as if they happen on much shorter time scales. It is not interacting with nearby NGC 4156 (barred spiral, 5′ NE, mag. 13.2).

NGC 4157

Constellation	Object type	RA, Dec	Approx. transit date at local midnight	Distance
Ursa Major	Barred spiral galaxy	12h 11.1 m, +50° 29′	April 18	60 million light years
Age	**Apparent size**	**Magnitude**	**Sky Atlas 2000.0 chart**	**Herald–Bobroff chart**
	6.7′ × 1.2′	11.4	2	C12

It is part of the Ursa Major galaxy cluster (see M 109/NGC 3992) as well as the M 109/NGC 3992 group of galaxies (see M 109/NGC 3992). This galaxy is nearly edge-on (its rotation axis is inclined about 80° from our line-of-sight). It has a mass of about one hundred billion Suns and an optical diameter of about 120 thousand light years.

NGC 4179

Constellation	Object type	RA, Dec	Approx. transit date at local midnight	Distance
Virgo	Lenticular galaxy	12h 12.9 m, +01° 18′	April 18	65 million light years
Age	**Apparent size**	**Magnitude**	**Sky Atlas 2000.0 chart**	**Herald–Bobroff chart**
	4.2′ × 1.3′	11.0	13	C48/D34

It is part of the NGC 4123 group of gravitationally bound galaxies that contains six galaxies whose other NGC members include NGC 4116 (barred spiral, 1° 56′ NW, mag. 12.0) and NGC 4123 (barred spiral, 1° 58′ NW, mag. 11.4).

NGC 4192 (M 98)

Constellation	Object type	RA, Dec	Approx. transit date at local midnight	Distance
Coma Berenices	Barred spiral galaxy	12h 13.8 m, +14° 54′	April 20	55 million light years
Age	**Apparent size**	**Magnitude**	**Sky Atlas 2000.0 chart**	**Herald–Bobroff chart**
	9.4′ × 2.3′	10.1	14	C48/D33/E10

M 98 is part of the Virgo galaxy cluster (see M 49/NGC 4472), near its western edge and thought to be lying toward the front of this cluster. It has a mass of about 2 hundred billion Suns and an optical diameter of about 150 thousand light years. It is nearly edge-on (with its axis of rotation inclined by about 80° from our line-of-sight). It has emission from its nucleus (perhaps from a LINER that is powered by both a starburst and accretion onto a massive central object – see M 81/NGC 3031).

NGC 4203

Constellation	Object type	RA, Dec	Approx. transit date at local midnight	Distance
Coma Berenices	Barred lenticular galaxy	12h 15.1 m, +33° 12'	April 19	40 million light years
Age	**Apparent size**	**Magnitude**	**Sky Atlas 2000.0 chart**	**Herald–Bobroff chart**
	3.5' × 3.2'	10.9	7	C30

This is a LINER galaxy (see M 81/NGC 3031) that is thought to belong to the class of LINERs which harbor a supermassive central black hole, having a mass of at most 6 million Suns in this galaxy. It is also relatively rich in atomic hydrogen gas for its galaxy type, perhaps having obtained this gas by accretion of another galaxy more than a billion years ago. It is part of the NGC 4274 galaxy group (see NGC 4274).

NGC 4214

Constellation	Object type	RA, Dec	Approx. transit date at local midnight	Distance
Canes Venatici	Irregular barred galaxy	12h 15.6 m, +36° 20'	April 19	15 million light years
Age	**Apparent size**	**Magnitude**	**Sky Atlas 2000.0 chart**	**Herald–Bobroff chart**
	8.4' × 7.2'	9.8	7	C30

This is a dwarf galaxy with intense star formation (a "starburst") in its central and southeast regions (0.5' SE) that are only a few million years old and contain thousands of hot young O-type stars, visible as bright knots in large amateur telescopes. The central starburst contains about 30 Wolf–Rayet stars (see NGC 2403 for explanation of "Wolf–Rayet" stars). The starbursts may have arisen because of a past merger or collision with another galaxy. It is part of the NGC 4631 galaxy group (see NGC 4631).

NGC 4216

Constellation	Object type	RA, Dec	Approx. transit date at local midnight	Distance
Virgo	Barred spiral galaxy	12h 15.9 m, +13° 09'	April 19	65 million light years
Age	**Apparent size**	**Magnitude**	**Sky Atlas 2000.0 chart**	**Herald–Bobroff chart**
	7.8' × 1.8'	10.0	13	C48/D33/E10

The nucleus of this galaxy has a different chemical make-up than the rest of the galaxy (the nucleus is richer in "metals" i.e. elements heavier than helium) and is also thought to be several billion years younger than its surrounding bulge. In addition, the nucleus (out to a radius of about 10") rotates independently of the rest of the galaxy (i.e. it is "kinematically decoupled"). This galaxy is nearly edge-on (with an inclination angle of about 70°) and has an optical diameter of about 150 thousand light years. It is part of the Virgo galaxy cluster (see M 49/NGC 4472), belonging to the M 87 clump of this cluster.

NGC 4244

Constellation	Object type	RA, Dec	Approx. transit date at local midnight	Distance
Canes Venatici	Spiral galaxy	12h 17.5 m, +37° 48′	April 20	15 million light years
Age	Apparent size	Magnitude	Sky Atlas 2000.0 chart	Herald–Bobroff chart
	15.9′ × 1.8′	10.4	7	C30

It is part of the NGC 4631 galaxy group (see NGC 4631). It has a mass of about 10 billion Suns, an optical diameter of about 70 thousand light years and is almost exactly edge-on.

NGC 4245

Constellation	Object type	RA, Dec	Approx. transit date at local midnight	Distance
Coma Berenices	Barred spiral galaxy	12h 17.6 m, +29° 36′	April 20	40 million light years
Age	Apparent size	Magnitude	Sky Atlas 2000.0 chart	Herald–Bobroff chart
	3.3′ × 2.4′	11.4	7	C30

It is part of the NGC 4274 galaxy group (see NGC 4274) and has a very low mass of atomic hydrogen (about 10 million Suns worth), perhaps due to tidal stripping in past interactions with other galaxies in this group.

NGC 4251

Constellation	Object type	RA, Dec	Approx. transit date at local midnight	Distance
Coma Berenices	Barred lenticular galaxy	12h 18.1 m, +28° 10′	April 20	40 million light years
Age	Apparent size	Magnitude	Sky Atlas 2000.0 chart	Herald–Bobroff chart
	3.6′ × 2.5′	10.7	7	C30

It is part of the NGC 4274 galaxy group (see NGC 4274). It has a diameter of about 40 thousand light years and is nearly edge-on. It has a very low mass of atomic and molecular hydrogen (less than 15 million Suns worth of each), which is typical of "early-type" galaxies like this lenticular galaxy.

NGC 4254 (M 99)

Constellation	Object type	RA, Dec	Approx. transit date at local midnight	Distance
Coma Berenices	Spiral galaxy	12h 18.8 m, +14° 25'	April 21	55 million light years

Age	Apparent size	Magnitude	Sky Atlas 2000.0 chart	Herald–Bobroff chart
	5.3' × 4.6'	9.9	14	C48/D33/E10

This is one of several galaxies whose nickname is the "Pinwheel Galaxy" (due to its multiple spiral arms evident in professional telescopic images). It is part of the Virgo galaxy cluster (see M 49/NGC 4472). It has a mass of a little over a hundred billion Suns and an optical diameter of about 90 thousand light years. Vigorous star formation is occurring throughout this galaxy. In professional telescopes, one of this galaxy's spiral arms is much more pronounced, possibly caused by a large cloud of gas (with a mass of hundreds of millions of Suns) that is falling into the galaxy and which may be the leftover scraps of a galaxy that M 99 previously disrupted.

NGC 4258 (M 106)

Constellation	Object type	RA, Dec	Approx. transit date at local midnight	Distance
Canes Venatici	Barred spiral galaxy	12h 19.0 m, +47° 18'	April 20	25 million light years

Age	Apparent size	Magnitude	Sky Atlas 2000.0 chart	Herald–Bobroff chart
	17.4' × 6.6'	8.4	7	C12/D35

This is a LINER galaxy (see M 81/NGC 3031) and is thought to harbor a central black hole with a mass of about 40 million Suns that is accreting a mass of perhaps 0.01 Suns per year. It is the nearest galaxy to us that has extragalactic astrophysical jets. These bipolar jets, driven by the central black hole, are thought to emanate from the nucleus at about a 30° angle from the plane of the disk and cause shocks in the gas of the disk which appear as "anomalous arms" in professional telescopic studies. The nucleus of this galaxy also contains water masers (see NGC 3079). This galaxy has an optical diameter of about 130 thousand light years. It is at the north end of the NGC 4258 (M 106) namesake group of 17 gravitationally bound galaxies that includes NGC 4144 (1° 45' WSW), NGC 4242 (1° 42' S), NGC 4460 (3° SE), NGC 4490 (6° SSE – see NGC 4490), NGC 4485 (6° SSE – see NGC 4485), NGC 4618 (7° SE – see NGC 4618), NGC 4625 (7° SE), NGC 4449 (3.5° SSE – see NGC 4449) and possibly NGC 4248 (14' WNW).

NGC 4261

Constellation	Object type	RA, Dec	Approx. transit date at local midnight	Distance
Virgo	Elliptical galaxy	12h 19.4 m, +05° 49'	April 20	130 million light years
Age	**Apparent size**	**Magnitude**	**Sky Atlas 2000.0 chart**	**Herald–Bobroff chart**
	3.8' × 3.5'	10.4	13	C48/D34/E12

This nearly edge-on galaxy has a central black hole with a mass of about 5 hundred million Suns, and is a LINER galaxy (see M 81/NGC 3031). It has a disk of dust and gas (with a diameter of 1.7", corresponding to about a thousand light years). This disk is accreting onto the nucleus, and may be left over from a single merger with another galaxy in the past. Professional telescopes find two jets pointing out from the nucleus perpendicular to this disk (one pointed E, the other pointed W) and extending for thousands of light years. NGC 4261 and nearby NGC 4264 (3.5' NE) are both thought to belong to the NGC 4261 group of gravitationally bound galaxies that contains about 30 galaxies and lies "behind" the Virgo galaxy cluster (see M 49/NGC 4472) at roughly twice the distance of the Virgo cluster.

NGC 4273

Constellation	Object type	RA, Dec	Approx. transit date at local midnight	Distance
Virgo	Barred spiral galaxy	12h 19.9 m, +05° 21'	April 20	130 million light years
Age	**Apparent size**	**Magnitude**	**Sky Atlas 2000.0 chart**	**Herald–Bobroff chart**
	2.3' × 1.5'	11.9	13	C48/D34/E12

It is part of the NGC 4261 galaxy group (see NGC 4261), as are its near neighbors, NGC 4277 (2' E) and NGC 4268 (4' SSW). Professional telescopes find NGC 4673 has a distorted shape whose cause remains unknown, but may be due to tidal distortion with another (unidentified) member of the NGC 4261 galaxy group within the past billion years or so.

NGC 4274

Constellation	Object type	RA, Dec	Approx. transit date at local midnight	Distance
Coma Berenices	Barred spiral galaxy	12h 19.8 m, +29° 37'	April 20	40 million light years
Age	**Apparent size**	**Magnitude**	**Sky Atlas 2000.0 chart**	**Herald–Bobroff chart**
	6.8' × 2.4'	10.4	7	C30

It is part of the NGC 4274 group of 17 gravitationally bound galaxies, that includes nearby NGC 4136 (2° 18' W), NGC 4173 (1° 41' WSW), NGC 4245 (28' W – see NGC 4245), NGC 4278 (20' S), NGC 4283 (19' SSE), NGC 4314 (40' ENE – see NGC 4314), NGC 4310 (42' SE), NGC 4251 (1° 30' SSW – see NGC 4251), NGC 4359 (2° 07' NNE), NGC 4020 (4° 36' W), NGC 4062 (4° NE), NGC 4525 (3° 06' ENE), and NGC 4203 (3° 43' NNW – see NGC 4203). A number of the galaxies in this group are deficient in atomic hydrogen (HI), and NGC 4274 is no exception – it is gas poor for its galaxy type, containing only about 2 hundred million Suns worth of atomic hydrogen. It has an optical diameter of about 80 thousand light years and is nearly edge-on (with an inclination angle of about 70°).

NGC 4278

Constellation	Object type	RA, Dec	Approx. transit date at local midnight	Distance
Coma Berenices	Elliptical galaxy	12h 20.1 m, +29° 17′	April 20	40 million light years
Age	**Apparent size**	**Magnitude**	**Sky Atlas 2000.0 chart**	**Herald–Bobroff chart**
	4.0′ × 3.9′	10.2	7	C30

The nucleus has strong emissions associated with photoionization so that it has been classified alternately as a Seyfert (see NGC 3227) or LINER galaxy (see M 81/NGC 3031). It is thought to contain a supermassive black hole at its center that has a mass of about 1.6 billion Suns. It is part of the NGC 4274 galaxy group (see NGC 4274), as is its close neighbor NGC 4283 (3′ NE). It is surprisingly rich in atomic hydrogen (HI) for an elliptical galaxy (containing hundreds of millions of solar masses of HI in a 5′ diameter ring) and contains about 250 globular clusters, both of which may be the result of this galaxy having swallowed a companion galaxy in the past.

NGC 4281

Constellation	Object type	RA, Dec	Approx. transit date at local midnight	Distance
Virgo	Lenticular galaxy	12h 20.4 m, +05° 23′	April 20	130 million light years
Age	**Apparent size**	**Magnitude**	**Sky Atlas 2000.0 chart**	**Herald–Bobroff chart**
	3.0′ × 1.6′	11.3	13	C48/D34/E12

Like its five immediate neighbors to the west (NGC 4273, NGC 4277, NGC 4268, NGC 4259 and NGC 4270), this galaxy belongs to the NGC 4261 galaxy group (see NGC 4261), which is roughly twice as far away from us as the Virgo galaxy cluster (see M 49/NGC 4472).

NGC 4293

Constellation	Object type	RA, Dec	Approx. transit date at local midnight	Distance
Coma Berenices	Barred lenticular galaxy	12h 21.2 m, +18° 23′	April 21	55 million light years
Age	**Apparent size**	**Magnitude**	**Sky Atlas 2000.0 chart**	**Herald–Bobroff chart**
	5.5′ × 2.9′	10.4	13	C30/D33/E10

This is a LINER galaxy (see M 81/NGC 3031). It is part of the Virgo galaxy cluster (see M 49/NGC 4472), lying in the M 87 clump of the Virgo cluster.

NGC (New General Catalogue) Objects

NGC 4303 (M 61)

Constellation	Object type	RA, Dec	Approx. transit date at local midnight	Distance
Virgo	Barred spiral galaxy	12h 21.9 m, +04° 28′	April 21	55 million light years
Age	**Apparent size**	**Magnitude**	**Sky Atlas 2000.0 chart**	**Herald–Bobroff chart**
	5.5′ × 2.9′	9.6	14	C48/D34/E12

This is one of fewer than about 10 galaxies that have had four or more recorded super-novae. It has a total mass of about 70 billion Suns, a diameter of about 90 thousand light years and is thought to be a Seyfert galaxy (see NGC 3227). It is part of the Virgo cluster (see M 49/NGC 4472) and is a grand-design spiral, meaning that it has two symmetrically placed spiral arms that extend over most its visible disk in professional telescopic images. Many giant, ionized hydrogen (star-forming) regions are present along its spiral arms, giving the arms an uneven brightness along their length in large amateur telescopes. It also has a massive circumnuclear star-forming disk (with a diameter of about 10″ and a mass of about 50 million Suns) that itself has a bar and spiral struc-ture. Total star-formation rates in this galaxy are probably 1–2 solar masses/year (about half that occurring in our Galaxy, but our Galaxy is about 10 times as massive). It is probably interacting with nearby NGC 4303A (10′ NE, mag. 13, listed as NGC 4303 on Herald–Bobroff chart E12) and NGC 4292 (12′ NW, mag. 12.2).

NGC 4314

Constellation	Object type	RA, Dec	Approx. transit date at local midnight	Distance
Coma Berenices	Barred spiral galaxy	12h 22.5 m, +29° 54′	April 21	40 million light years
Age	**Apparent size**	**Magnitude**	**Sky Atlas 2000.0 chart**	**Herald–Bobroff chart**
	3.9′ × 3.7′	10.6	7	C30

It is part of the NGC 4274 galaxy group (see NGC 4274). It is nearly face-on (with an inclination angle of 23°). Professional telescopes find a ring of star formation around the nucleus with a diameter of about 16″. This ring is the only place star formation is currently occurring in this galaxy, and has been going on for a few tens of millions of years. The ring is believed to be related to the presence of a nuclear bar (with an oval shape and a diameter of 8″ in professional telescopic studies). This bar is thought to have driven gas from the disk inward into a ring situated between two so-called "inner Lindblad resonances" (where the speed of density waves resonantly amplifies oscilla-tions in the orbits of matter in the disk). This is a LINER galaxy (see M 81/NGC 3031) and is believed to harbor a supermassive central black hole. It has an optical diameter of about 45 thousand light years.

NGC 4321 (M 100)

Constellation	Object type	RA, Dec	Approx. transit date at local midnight	Distance
Coma Berenices	Barred spiral galaxy	12h 22.9 m, +15° 49'	April 22	55 million light years
Age	**Apparent size**	**Magnitude**	**Sky Atlas 2000.0 chart**	**Herald–Bobroff chart**
	7.5' × 6.1'	9.4	14	C48/D33/E10

It is part of the Virgo galaxy cluster (see M 49/NGC 4472), being the brightest and largest spiral galaxy in the Virgo cluster. It is a "grand-design spiral", meaning that it has two symmetrically placed spiral arms that extend over most of its visible disk in professional telescopic images. Resonance associated with the bar and the "Lindblad resonances" (where the speed of density waves resonantly amplifies oscillations in the orbits of matter in the disk) are thought to be triggering a ring of star formation (a "starburst") in the central region of M 100. This ring has a radius of 7.5–20" and is perhaps 10 million years old. M 100 is one of fewer than 10 galaxies that has had four or more recorded supernovae. M 100 has a mass of about 2 hundred billion Suns and an optical diameter of about 120 thousand light years.

NGC 4346

Constellation	Object type	RA, Dec	Approx. transit date at local midnight	Distance
Canes Venatici	Barred lenticular galaxy	12h 23.5 m, +47° 00'	April 21	60 million light years
Age	**Apparent size**	**Magnitude**	**Sky Atlas 2000.0 chart**	**Herald–Bobroff chart**
	3.2' × 1.3'	11.1	7	C12/D35

It is part of the Ursa Major galaxy cluster (see M 109/NGC 3992), as well as the NGC 4051 galaxy group (see NGC 4051).

NGC 4350

Constellation	Object type	RA, Dec	Approx. transit date at local midnight	Distance
Coma Berenices	Lenticular galaxy	12h 24.0 m, +16° 42'	April 21	55 million light years
Age	**Apparent size**	**Magnitude**	**Sky Atlas 2000.0 chart**	**Herald–Bobroff chart**
	2.9' × 1.6'	11.0	13	C30/D33/E10

It contains a massive central dark object (probably a black hole) with a mass of several hundred million Suns. The disk of this galaxy accounts for a little more than a quarter of the light it emits, while its diffuse ellipsoidal part accounts for a little less than 3/4 of its light. The mass of this galaxy is about 50 billion Suns. It is part of the M 87 clump of the Virgo galaxy cluster (see M 49/NGC 4472).

NGC 4361

Constellation	Object type	RA, Dec	Approx. transit date at local midnight	Distance
Corvus	Planetary nebula	12h 24.5 m, –18° 47′	April 21	4 thousand light years
Age	Apparent size	Magnitude	Sky Atlas 2000.0 chart	Herald–Bobroff chart
	1′	10.9	13	C66

Professional telescopic studies suggest that this nebula is one of only seven known planetary nebula that has a quadrapolar structure i.e. it has two pairs of bipolar lobes, one pair at position angles of 105° and 285° with a diameter of 68″, while the other pair is at position angles of 25° and 225° and a diameter of 90″. These lobes may have been produced by two separate mass ejection events (about a thousand years apart) with each event producing one of the bipolar lobes by the same mechanism that produces the more common bipolar planetary nebulae (see M 27/NGC 6853). The bipolar lobes may have different orientations because the circumstellar toroidal ring of material that leads to the formation of the jets (see M 27/NGC 6853) was oriented in different directions (possibly due to precession of this ring due to the presence of a secondary star i.e. the central star may be a binary star). These lobes (which are about 5 thousand years old) are thought to be embedded in a spheroidal shell with a diameter of about 50″ (that is about 10 thousand years old), all of which is surrounded by a filamentary outer halo (nearly 2′ in diameter) that has regions of high expansion velocity (70 km/s). The magnitude 13 central star is visible in amateur scopes and is thought to be a Population II star (i.e. it formed from the original, very old, material in our Galaxy).

NGC 4365

Constellation	Object type	RA, Dec	Approx. transit date at local midnight	Distance
Virgo	Elliptical galaxy	12h 24.5 m, +07° 19′	April 21	70 million light years
Age	Apparent size	Magnitude	Sky Atlas 2000.0 chart	Herald–Bobroff chart
	6.5′ × 4.9′	9.6	13	C48/D34/E12

This is one of about 30% of elliptical and lenticular galaxies that has a "decoupled core", meaning that the core of the galaxy rotates about a different axis than the rest of the galaxy. In this galaxy, the inner core (within radii of 2–3″) rotates about the minor axis of the elliptical shape this galaxy makes in the sky, while the rest of the galaxy rotates about the major axis (i.e. at right angles to the rotation axis of the inner core). This is thought to be the result of previous accretion of material in one or more galaxy mergers, which in this galaxy are believed to have ended many billions of years ago. This galaxy is thought to lie beyond the main Virgo galaxy cluster (see M 49/NGC 4472).

NGC 4371

Constellation	Object type	RA, Dec	Approx. transit date at local midnight	Distance
Virgo	Barred lenticular galaxy	12h 24.9 m, +11° 42′	April 21	55 million light years
Age	**Apparent size**	**Magnitude**	**Sky Atlas 2000.0 chart**	**Herald–Bobroff chart**
	4.0′ × 2.3′	10.8	13	C48/D33/E10

It is part of the Virgo galaxy cluster (see M 49/NGC 4472). The nucleus of this galaxy is not active, and has a ring (21″ in diameter) around it in professional telescopic studies.

NGC 4374 (M 84)

Constellation	Object type	RA, Dec	Approx. transit date at local midnight	Distance
Virgo	Elliptical galaxy	12h 25.1m, +12° 53′	April 23	55 million light years
Age	**Apparent size**	**Magnitude**	**Sky Atlas 2000.0 chart**	**Herald–Bobroff chart**
	6.7′ × 6.0′	9.1	14	C48/D33/E10

It is part of the Virgo galaxy cluster (see M 49/NGC 4472) and a physically close companion to M 86. M 84 lies at one end of the "Markarian Chain" of galaxies that lies along a "chain" NE of M 84 and includes eight galaxies: M 84 (NGC 4374), M 86 (NGC 4406), NGC 4435, NGC 4438, NGC 4461, NGC 4458, NGC 4473 and NGC 4477. These galaxies are moving like a rigid, tumbling chain thrown away from Earth at several hundred km/s with the chain tumbling so that the W side of the chain (M 84) is actually moving toward us while the E side (NGC 4477) is moving doubly fast away from us. The nucleus of M 84 is an AGN (or "active galactic nucleus"), in which energetic emission is caused by accretion onto a massive central object, which for M 84 is thought to be a black hole with a mass of several hundred million Suns. This AGN powers the two jets and associated lobes that are evident in the nuclear region (the inner few arcseconds) in professional telescopes at radio wavelengths, the radio waves being due to synchrotron emission of high-speed electrons gyrating wildly in a magnetic field. M 84 is one of fewer than 20 galaxies to have had three or more recorded supernovae.

NGC 4382 (M 85)

Constellation	Object type	RA, Dec	Approx. transit date at local midnight	Distance
Coma Berenices	Lenticular galaxy	12h 25.4m, +18° 11′	April 23	55 million light years
Age	**Apparent size**	**Magnitude**	**Sky Atlas 2000.0 chart**	**Herald–Bobroff chart**
	7.4′ × 5.9′	9.1	14	C30/D33/E10

M 85 is at the northern edge of the Virgo galaxy cluster (see M 49/NGC 4472) and a close physical companion to NGC 4394 (7′ ENE). It is viewed nearly face-on and is unusual for a lenticular galaxy because it contains young stars (i.e. less than a few billion years old) in its inner regions, which may have arisen from a past merger with another galaxy or because of interaction with nearby NGC 4394.

NGC 4388

Constellation	Object type	RA, Dec	Approx. transit date at local midnight	Distance
Virgo	Spiral galaxy	12h 25.8 m, +12° 40′	April 21	55 million light years
Age	**Apparent size**	**Magnitude**	**Sky Atlas 2000.0 chart**	**Herald–Bobroff chart**
	5.6′ × 1.5′	11.0	13	C48/D33/E10

It is part of the Virgo galaxy cluster (see M 49/NGC 4472). It has a mass of about 90 billion Suns and an optical diameter of about 90 thousand light years. It is a Seyfert galaxy in which emission from the nucleus is thought to occur due to accretion of matter onto a massive central black hole. The galaxy is nearly edge-on (with an inclination angle of 72°).

NGC 4394

Constellation	Object type	RA, Dec	Approx. transit date at local midnight	Distance
Coma Berenices	Barred spiral galaxy	12h 25.9 m, +18° 13′	April 22	55 million light years
Age	**Apparent size**	**Magnitude**	**Sky Atlas 2000.0 chart**	**Herald–Bobroff chart**
	3.4′ × 3.2′	10.9	13	C30/D33/E10

It is part of the M 87 clump of the Virgo cluster (see M 49/NGC 4472), as is nearby M 85 (7′ W – see M 85/NGC 4382). It lies at the northern boundary of the Virgo cluster. NGC 4394 is a LINER ("low-ionization nuclear emission region") galaxy (see M 81/NGC 3031). It has a diameter of about 50 thousand light years and is nearly face-on (its inclination angle is about 22°, meaning that it rotates about an axis that is tilted from our line-of-sight by about 22°).

NGC 4406 (M 86)

Constellation	Object type	RA, Dec	Approx. transit date at local midnight	Distance
Virgo	Elliptical galaxy	12h 26.2m, +12° 57′	April 23	55 million light years
Age	Apparent size	Magnitude	Sky Atlas 2000.0 chart	Herald–Bobroff chart
	9.8′ × 6.3′	8.9	14	C48/D33/E10

It is part of the Virgo galaxy cluster (see M 49/NGC 4472), lying toward the back side of the main concentration of the Virgo cluster (i.e. lying perhaps a little less than 10 million light years farther away than M 87). Along with its close physical companion M 84, it is part of the "Markarian Chain" of galaxies (see M 84/NGC 4374). M 86 is moving rapidly toward us compared to the rest of the Virgo cluster, resulting in M 86 having a high speed (over a thousand km/s) relative to the material between galaxies in the cluster (the "intracluster medium"). This galaxy's high-speed movement through the intracluster medium is thought to be causing gas to be stripped from the galaxy (so-called "ram-pressure stripping").

NGC 4414

Constellation	Object type	RA, Dec	Approx. transit date at local midnight	Distance
Coma Berenices	Spiral galaxy	12h 26.5 m, +31° 13′	April 22	65 million light years
Age	Apparent size	Magnitude	Sky Atlas 2000.0 chart	Herald–Bobroff chart
	3.6′ × 2.0′	10.1	7	C30

This is a flocculent spiral galaxy (see NGC 3521 for explanation). It contains an unusually large surface density of neutral gas in its disk (about 10 times that of the Milky Way), which would normally give intense star formation if gravitationally disturbed by interaction with another galaxy. The relative isolation of NGC 4414 is thought to have kept this from occurring to date.

NGC 4419

Constellation	Object type	RA, Dec	Approx. transit date at local midnight	Distance
Coma Berenices	Barred spiral galaxy	12h 26.9 m, +15° 03′	April 22	55 million light years
Age	Apparent size	Magnitude	Sky Atlas 2000.0 chart	Herald–Bobroff chart
	3.3′ × 1.2′	11.2	13	C30/D33/E10

It is part of the M 87 clump of the Virgo cluster (see M 49/NGC 4472). This is a LINER galaxy (see M 81/NGC 3031) with active star formation thought to be occurring in the nucleus and circumnuclear disk. Like many of the spirals in the Virgo cluster, this galaxy has much less atomic hydrogen gas than average for spirals in general. This is thought to be caused by stripping of this gas from the outer regions of the galaxy due to the galaxy's extreme speed (1.3 thousand km/s for this galaxy) relative to the material between galaxies in the cluster (the "intracluster medium"). The galaxy has a dynamical mass of about 90 billion Suns, an optical diameter of about 50 thousand light years, and an inclination angle of about 70° (i.e. it is nearly edge-on).

NGC 4429

Constellation	Object type	RA, Dec	Approx. transit date at local midnight	Distance
Virgo	Lenticular galaxy	12h 27.4 m, +11° 06′	April 22	55 million light years
Age	**Apparent size**	**Magnitude**	**Sky Atlas 2000.0 chart**	**Herald–Bobroff chart**
	5.8′ × 2.8′	10.0	13	C48/D33/E10

It is part of the Virgo galaxy cluster (see M 49/NGC 4472). This is a LINER galaxy (see M 81/NGC 3031). It has an optical diameter of about 90 thousand light years and an inclination angle of about 75°. Professional telescopic studies find that the nucleus of this galaxy (radius of 1″) is a few billion years younger and chemically distinct from the remainder of the galaxy, in addition to having a nuclear ring (radius 4–6″) from a previous starburst ring. These structures may have been caused by orbital resonant interactions, but this remains uncertain.

NGC 4435

Constellation	Object type	RA, Dec	Approx. transit date at local midnight	Distance
Virgo	Barred lenticular galaxy	12h 27.7 m, +13° 05′	April 22	55 million light years
Age	**Apparent size**	**Magnitude**	**Sky Atlas 2000.0 chart**	**Herald–Bobroff chart**
	3.0′ × 2.2′	10.8	13	D33/E10

It has been suggested that NGC 4435 is the galaxy that was involved in the high-speed collision that disturbed nearby NGC 4438 (4′ SSE – see NGC 4438), although this is uncertain and some other unidentified galaxy may have been the culprit. A central black hole with a mass of a billion Suns is thought to be present in NGC 4435. This galaxy is edge-on (with an inclination angle of 90°) and has a mass of about 80 billion Suns. It is part of the "Markarian Chain" of galaxies (see NGC 4473).

NGC 4438

Constellation	Object type	RA, Dec	Approx. transit date at local midnight	Distance
Virgo	Lenticular galaxy	12h 27.8 m, +13° 00′	April 22	55 million light years
Age	**Apparent size**	**Magnitude**	**Sky Atlas 2000.0 chart**	**Herald–Bobroff chart**
	8.5′ × 3.0′	10.2	13	C48/D33/E10

This is the most disrupted galaxy in the Virgo galaxy cluster. This disruption is thought to be the result of a high velocity (1 thousand km/s) collision with another galaxy (see NGC 4435) some hundred million years ago (having a closest approach of a few tens of thousands of light years). Filaments of interstellar material that were extruded in this collision are now thought to be falling back onto NGC 4438. This galaxy has a mass of about 4 hundred billion Suns and is a LINER galaxy (see M 81/NGC 3031). It is part of the "Markarian Chain" of galaxies (see NGC 4473).

NGC 4442

Constellation	Object type	RA, Dec	Approx. transit date at local midnight	Distance
Virgo	Barred lenticular galaxy	12h 28.1 m, +09° 48′	April 22	55 million light years
Age	Apparent size	Magnitude	Sky Atlas 2000.0 chart	Herald–Bobroff chart
	4.5′ × 1.8′	10.4	13	C48/D33/E12

It is part of the Virgo cluster (see M 49/NGC 4472) as well as the NGC 4442 galaxy group of five galaxies the brightest of which are NGC 4469 (1° 6′ SSE, mag. 11.2) and IC 3521 (3° SSE, mag. 13). It has a mass of about 50 billion Suns and is nearly edge-on (with an inclination angle of about 70°).

NGC 4448

Constellation	Object type	RA, Dec	Approx. transit date at local midnight	Distance
Coma Berenices	Barred spiral galaxy	12h 28.3 m, +28° 37′	April 22	30 million light years
Age	Apparent size	Magnitude	Sky Atlas 2000.0 chart	Herald–Bobroff chart
	3.6′ × 1.3′	11.1	7	C30

It is nearly edge-on (with an inclination angle of about 70°). It has an optical diameter of about 30 thousand light years and has an outer ring. Outer rings require a galaxy to avoid tidal interactions with other galaxies (since such interactions destroy these rings). As a result, galaxies with rings do not usually belong to a gravitationally bound group, and this galaxy is no exception.

NGC 4449

Constellation	Object type	RA, Dec	Approx. transit date at local midnight	Distance
Canes Venatici	Irregular barred galaxy	12h 28.2 m, +44° 06′	April 22	13 million light years
Age	Apparent size	Magnitude	Sky Atlas 2000.0 chart	Herald–Bobroff chart
	6.2′ × 4.9′	9.6	7	C30/D35

It has an optical diameter of about 20 thousand light years. Active star formation is occurring along its bar and it has a starburst nucleus. The nucleus contains a star cluster with several hundred thousand hot, young stars that are a few million years old, as well as over 250 giant ionized hydrogen (HII) regions (star-forming regions like M 42). Several of these HII regions are visible as bright knots in large amateur telescopes. Professional telescopes find it has a large atomic hydrogen (HI) region outside the optically visible region, that extends out about 10 times the optical diameter of this galaxy and has a mass of about a billion Suns. The origin of this HI gas is not known, but it may be the result of a close encounter with M 94 (NGC 4736, 5° ESE) several billion years ago that left a hundred thousand light year diameter disk of HI gas around NGC 4449. Since then, interaction with nearby UGC 7577 (irregular dwarf galaxy, mag. 12, 37′ S) has redistributed this gas into its current lop-sided arm, streamers, and disk. NGC 4449 has a mass of about 3 billion Suns within its optical diameter, while UGC 7577 is about 20% the mass of NGC 4449. NGC 4449 is part of the M 106/NGC 4258 galaxy group.

NGC 4450

Constellation	Object type	RA, Dec	Approx. transit date at local midnight	Distance
Coma Berenices	Spiral galaxy	12h 28.5 m, +17° 05′	April 22	55 million light years
Age	**Apparent size**	**Magnitude**	**Sky Atlas 2000.0 chart**	**Herald–Bobroff chart**
	5.4′ × 4.1′	10.1	13	C30/D33/E10

It is part of the Virgo galaxy cluster (see M 49/NGC 4472). This is a LINER (low-ionization nuclear emission region) galaxy (see M 81/NGC 3031), which in this case is thought to be caused by photoionization due to accretion onto a super-massive central black hole. This galaxy has an optical diameter of about 90 thousand light years.

NGC 4459

Constellation	Object type	RA, Dec	Approx. transit date at local midnight	Distance
Coma Berenices	Lenticular galaxy	12h 29.0 m, +13° 59′	April 23	55 million light years
Age	**Apparent size**	**Magnitude**	**Sky Atlas 2000.0 chart**	**Herald–Bobroff chart**
	4.0′ × 3.1′	10.4	13	C30/D33/E10

This galaxy is thought to contain a supermassive central black hole with a mass of about a billion Suns. It is part of the M 87 clump of the Virgo galaxy cluster (see M 49/NGC 4472).

NGC 4472 (M 49)

Constellation	Object type	RA, Dec	Approx. transit date at local midnight	Distance
Virgo	Elliptical galaxy	12h 29.8 m, +08° 00′	April 23	55 million light years
Age	**Apparent size**	**Magnitude**	**Sky Atlas 2000.0 chart**	**Herald–Bobroff chart**
	9.8′ × 8.2′	8.4	14	C48/D34/E11

This is a giant elliptical galaxy and the brightest member of the Virgo galaxy cluster, which is the nearest galaxy cluster to us. Galaxy clusters are the largest stable structures in the Universe, although only about 1% of galaxies belong to galaxy clusters. The Virgo galaxy cluster contains many hundreds of galaxies, which are bunched into several subclusters with M 87 and M 49 being principal members of the two main subclusters. M 49 lies perhaps a few million light years closer than M 87. Occupying a roughly rectangular shape in the sky (8° E–W × 16° N–S), the major concentration of the Virgo cluster extends several tens of millions of light years around the cluster center near M 87 (but, for simplicity, all its members are listed herein as being at the same distance). M 49 is thought to be falling toward the center of the Virgo cluster at about a thousand km/s. Much of the mass in the Virgo cluster is dark matter, with the total mass of the Virgo cluster being many hundreds of trillions of solar masses. M 49 contains about 6 thousand globular clusters of two different ages, which suggests M 49 may be the product of a merger event in its past. The center of M 49 is thought to contain a supermassive black hole with a mass of about 1/2 billion Suns. The luminous mass of M 49 is roughly 2 hundred billion Suns, but its total mass (including dark matter in its halo) may be almost 2 trillion Suns. Its optical diameter is about 150 thousand light years.

NGC 4473

Constellation	Object type	RA, Dec	Approx. transit date at local midnight	Distance
Coma Berenices	Elliptical galaxy	12h 29.8 m, +13° 26′	April 23	55 million light years
Age	**Apparent size**	**Magnitude**	**Sky Atlas 2000.0 chart**	**Herald–Bobroff chart**
	4.2′ × 2.6′	10.2	13	C48/D33/E10

It contains a massive central black hole. The inner 45″ radius of this galaxy has a mass of about 50 billion Suns. This galaxy is rotating slowly (60 km/s), with the west side approaching and the east side receding with respect to the galaxy's center. It is part of the M 87 clump of the Virgo galaxy cluster (see M 49/NGC 4472), as well as being part of the "Markarian Chain" of galaxies that lies along a "chain" NE of M 84 that includes eight galaxies: M 84 (NGC 4374), M 86 (NGC 4406), NGC 4435, NGC 4438, NGC 4461, NGC 4458, NGC 4473 and NGC 4477. These galaxies are moving like a rigid, tumbling chain thrown away from Earth at several hundred km/s with the chain tumbling so that the W side of the chain (M 84) is actually moving toward us while the E side (NGC 4477) is moving doubly fast away from us.

NGC 4477

Constellation	Object type	RA, Dec	Approx. transit date at local midnight	Distance
Coma Berenices	Barred lenticular galaxy	12h 30.0 m, +13° 38′	April 23	55 million light years
Age	**Apparent size**	**Magnitude**	**Sky Atlas 2000.0 chart**	**Herald–Bobroff chart**
	3.7′ × 3.3′	10.4	13	C48/D33/E10

It is part of the M 87 clump of the Virgo galaxy cluster (see M 49/NGC 4472), as well as the "Markarian Chain" of galaxies (see NGC 4473) of which it forms the eastern end. It is a Seyfert galaxy (where a supermassive object in this galaxy's center accumulates nearby material and causes strong nuclear emission).

NGC 4478

Constellation	Object type	RA, Dec	Approx. transit date at local midnight	Distance
Virgo	Elliptical galaxy	12h 30.3 m, +12° 20′	April 23	55 million light years
Age	**Apparent size**	**Magnitude**	**Sky Atlas 2000.0 chart**	**Herald–Bobroff chart**
	1.8′ × 1.5′	11.4	13	C48/D33/E09

This galaxy is part of the Virgo galaxy cluster (see M 49/NGC 4472) and is thought to be a companion of M 87 (see M 87/NGC 4486, 8′ ENE), as is its neighbor NGC 4476 (4.5′ WNW). NGC 4478 has suffered in its companionship with M 87 though, since it has lost material to M 87, with some of the many globular clusters in M 87 believed to have been stolen from NGC 4478.

NGC 4485

Constellation	Object type	RA, Dec	Approx. transit date at local midnight	Distance
Canes Venatici	Irregular barred galaxy	12h 30.5 m, +41° 42′	April 23	30 million light years
Age	**Apparent size**	**Magnitude**	**Sky Atlas 2000.0 chart**	**Herald–Bobroff chart**
	2.4′ × 1.8′	11.9	7	C30

This galaxy is interacting with nearby NGC 4490 (4′ SSE – see NGC 4490), with their closest approach a few hundred million years ago. Indeed, much of the interstellar medium in NGC 4485 is thought to have been stripped out of NGC 4485 because this galaxy has been passing through the atomic hydrogen (HI) gas halo of NGC 4490 (so-called "ram-pressure" stripping). Professional telescopes find a large HI cloud SW of NGC 4485 (extending in a bridge between the two galaxies due to their tidal interaction) and that probably used to be part of NGC 4485, but was blown out of this galaxy by ram-pressure stripping. NGC 4485 has a diameter of about 20 thousand light years and is about 4 times less massive than NGC 4490. NGC 4485 is part of the M 106 galaxy group (see M 106/NGC 4258).

NGC 4486 (M 87)

Constellation	Object type	RA, Dec	Approx. transit date at local midnight	Distance
Virgo	Elliptical galaxy	12h 30.8m, +12° 23′	April 24	55 million light years
Age	**Apparent size**	**Magnitude**	**Sky Atlas 2000.0 chart**	**Herald–Bobroff chart**
	8.7′ × 6.6′	8.6	14	C48/D33/E09

Lying near the center of the Virgo galaxy cluster (see M 49/NGC 4472), this is a heavy-weight in the world of galaxies. Its exact mass is uncertain but is probably several trillion solar masses (although much of this is dark matter). This is many hundreds of times more massive than the average galaxy. M 87 lies at the heart of the largest of the two major subclusters within the Virgo cluster (the other being associated with M 49/NCG 4472), with the M 87 clump of the Virgo cluster having a mass of several hundred trillion Suns. The center of M 87 is thought to contain a supermassive black hole with a mass of several billion Suns. This black hole is believed to be what drives the active galactic nucleus (in which energetic emission is caused by accretion onto the massive central black hole) that gives rise to the optically one-sided jet in M 87. This jet is the major part of the radio source known as Virgo A, the radio waves being due to synchrotron emission of high-speed electrons gyrating wildly in a magnetic field. This jet extends for thousands of light years (its brightest parts extending out about 25″), with material in the jet traveling at significant fractions of the speed of light. M 87 has one of the largest numbers of globular clusters of any galaxy, with more than 13 thousand (compare to the Milky Way's 150 or so). The optical diameter of M 87 is about 140 thousand light years (which is more than twice the diameter of the average galaxy), but it extends out several times this distance in professional telescopic studies.

NGC 4490

Constellation	Object type	RA, Dec	Approx. transit date at local midnight	Distance
Canes Venatici	Barred spiral galaxy	12h 30.6 m, +41° 39′	April 23	30 million light years
Age	**Apparent size**	**Magnitude**	**Sky Atlas 2000.0 chart**	**Herald–Bobroff chart**
	6.4′ × 3.2′	9.8	7	C30

Intense star formation (nearly 5 new Suns worth per year) is occurring throughout the disk of NGC 4490, an unusual situation, since such starbursts are normally confined to the nuclear regions of galaxies. It has been proposed that this may be due to NGC 4490 being a young galaxy (perhaps 2 billion years old) still in its active star-formation stage. The atomic hydrogen (HI) envelope (visible in professional telescopic studies) surrounding this galaxy and nearby NGC 4485 (4′ NNW – see NGC 4485) is one of the largest known, extending out nearly 10 times the optical diameter of NGC 4490 and having a mass of nearly 20 billion Suns. This galaxy is part of the M 106 galaxy group (see M 106/NGC 4258).

NGC 4494

Constellation	Object type	RA, Dec	Approx. transit date at local midnight	Distance
Coma Berenices	Elliptical galaxy	12h 31.4 m, +24° 46′	April 23	40 million light years
Age	**Apparent size**	**Magnitude**	**Sky Atlas 2000.0 chart**	**Herald–Bobroff chart**
	4.5′ × 4.3′	9.8	7	C30

It is part of a gravitationally bound galaxy triplet with NGC 4565 (1° 07′ E – see NGC 4565) and NGC 4562 (barred spiral, 57′ E, mag. 13.5). It is a LINER ("low-ionization nuclear emission region") galaxy (see M 81/NGC 3031).

NGC 4501 (M 88)

Constellation	Object type	RA, Dec	Approx. transit date at local midnight	Distance
Coma Berenices	Spiral galaxy	12h 32.0m, +14° 25′	April 24	55 million light years
Age	**Apparent size**	**Magnitude**	**Sky Atlas 2000.0 chart**	**Herald–Bobroff chart**
	6.8′ × 3.7′	9.6	14	C48/D33/E09

M 88 is part of the Virgo galaxy cluster (see M 49/NGC 4472) and is a Seyfert galaxy (where a supermassive object in this galaxy's center accumulates nearby gas resulting in strong emission from the nucleus). It has a mass of about 1/4 billion Suns and an optical diameter of about 110 thousand light years. The nucleus of this galaxy (within a radius of about 4″) has a different chemical makeup than the rest of the galaxy. The galaxy is a "flocculent" spiral, meaning that it lacks any obvious azimuthally symmetric spiral arm pattern (see NGC 3521).

NGC 4517

Constellation	Object type	RA, Dec	Approx. transit date at local midnight	Distance
Virgo	Spiral galaxy	12h 32.8 m, +00° 07′	April 23	55 million light years
Age	Apparent size	Magnitude	Sky Atlas 2000.0 chart	Herald–Bobroff chart
	10.2′ × 1.7′	10.4	14	C48/D34

It is part of the Virgo galaxy cluster (see M 49/NGC 4472), thought to be lying toward the near side of this cluster. It is nearly edge-on and has a mass of about 45 billion Suns. NGC 4437 is a duplicate entry of NGC 4517.

NGC 4526

Constellation	Object type	RA, Dec	Approx. transit date at local midnight	Distance
Virgo	Barred lenticular galaxy	12h 34.0 m, +07° 42′	April 24	55 million light years
Age	Apparent size	Magnitude	Sky Atlas 2000.0 chart	Herald–Bobroff chart
	7.0′ × 2.5′	9.7	14	C48/D34/E11

This galaxy is part of the Virgo galaxy cluster (see M 49/NGC 4472), and lies near the core of the cluster.

NGC 4527

Constellation	Object type	RA, Dec	Approx. transit date at local midnight	Distance
Virgo	Barred spiral galaxy	12h 34.1 m, +02° 39′	April 24	55 million light years
Age	Apparent size	Magnitude	Sky Atlas 2000.0 chart	Herald–Bobroff chart
	5.9′ × 2.3′	10.5	14	C48/D34/E11

This galaxy is part of the Virgo galaxy cluster (see M 49/NGC 4227), and lies toward the closer side of this cluster. A massive dark core is present in this galaxy, with nearly 10 billion solar masses present in the inner 10″ radius (corresponding to a radius of about 3 thousand light years). Most of the galaxy rotates at about 2 hundred km/s about an axis that is inclined 69° from our line-of-sight. This galaxy has an optical diameter of about 90 thousand light years.

NGC 4535

Constellation	Object type	RA, Dec	Approx. transit date at local midnight	Distance
Virgo	Barred spiral galaxy	12h 34.3 m, +08° 12′	April 24	55 million light years
Age	Apparent size	Magnitude	Sky Atlas 2000.0 chart	Herald–Bobroff chart
	6.9′ × 5.4′	10.0	14	C48/D34/E11

It is part of the Virgo galaxy cluster (see M 49/NGC 4472). It has an optical diameter of about 110 thousand light years. Its rotation axis is inclined at 47° from our line-of-sight.

NGC 4536

Constellation	Object type	RA, Dec	Approx. transit date at local midnight	Distance
Virgo	Barred spiral galaxy	12h 34.4 m, +02° 11′	April 24	55 million light years
Age	Apparent size	Magnitude	Sky Atlas 2000.0 chart	Herald–Bobroff chart
	7.1′ × 3.1′	10.6	14	C48/D34/E11

It is part of the Virgo galaxy cluster (see M 49/NGC 4427). This galaxy is nearly edge-on (with an inclination angle of about 65°). It is a starburst galaxy, meaning that intense star formation is occurring in its central regions. However, it is thought that star formation has come to an end in the nucleus (so that supernovae power the emission there) but star formation continues in a circumnuclear ring (about 7″ in diameter).

NGC 4546

Constellation	Object type	RA, Dec	Approx. transit date at local midnight	Distance
Virgo	Barred lenticular galaxy	12h 35.5 m, −03° 48′	April 24	60 million light years
Age	Apparent size	Magnitude	Sky Atlas 2000.0 chart	Herald–Bobroff chart
	3.3′ × 1.6′	10.3	14	C47/D31

This galaxy is unusual in that it contains a gas disk that counter-rotates (at several hundred km/s) relative to the stars in the disk. The gas disk has a radius of about 25 thousand light years and contains several hundred million solar masses. The gas is thought to be left over from an accreted gas-rich dwarf galaxy. The galaxy is nearly edge-on (inclination angle of about 70°). The diameter of this galaxy is about 60 thousand light years.

NGC 4548 (M 91)

Constellation	Object type	RA, Dec	Approx. transit date at local midnight	Distance
Coma Berenices	Barred spiral galaxy	12h 35.4 m, +14° 30′	April 24	55 million light years
Age	Apparent size	Magnitude	Sky Atlas 2000.0 chart	Herald–Bobroff chart
	5.2′ × 4.2′	10.1	14	C48/D33/E09

This galaxy is part of the M 87 clump of the Virgo galaxy cluster (see M 49/NGC 4472). It is a LINER galaxy (see M 81/NGC 3031). Like many spirals in galaxy clusters, this galaxy has much less atomic hydrogen gas than average for spirals in general. This is thought to be caused by stripping of this gas (so-called "ram-pressure stripping") from the galaxy due to the galaxy's motion relative to the material between galaxies in the cluster (the "intracluster medium"). Most of this galaxy rotates at about 250 km/s about an axis inclined to our line-of-sight by 40°. It is thought that a mistake was made by Messier in listing the position of this object, leading to confusion as to which object M 91 referred to, until the probable origin of Messier's error was uncovered in 1969 by W.C. Williams.

NGC 4550

Constellation	Object type	RA, Dec	Approx. transit date at local midnight	Distance
Virgo	Barred lenticular galaxy	12h 35.5 m, +12° 13′	April 24	55 million light years
Age	**Apparent size**	**Magnitude**	**Sky Atlas 2000.0 chart**	**Herald–Bobroff chart**
	3.3′ × 0.9′	11.7	14	C48/D33/E09

It is part of the Virgo galaxy cluster (see M 49/NGC 4472). This galaxy has a stellar mass of about 20 billion Suns. It is one of only two known galaxies that consists of two large, counter-rotating stellar disks (the other being NGC 7217 – see NGC 7217). In NGC 4550 the two disks have nearly equal masses and may be the result of the merger of two galaxies with opposite directions of rotation, with this event having occurred at least several billions of years ago. The presence of about 10 million Suns worth of molecular gas in this galaxy is thought to be the result of accretion of a dwarf galaxy within the past few hundred million years or so.

NGC 4552 (M 89)

Constellation	Object type	RA, Dec	Approx. transit date at local midnight	Distance
Virgo	Elliptical galaxy	12h 35.7m, +12° 33′	April 24	55 million light years
Age	**Apparent size**	**Magnitude**	**Sky Atlas 2000.0 chart**	**Herald–Bobroff chart**
	5.3′ × 4.8′	9.8	14	C48/D33/E09

It is part of the Virgo galaxy cluster (see M 49/NGC 4472). A supermassive object in this galaxy's center (thought to be a black hole with a mass of many hundreds of millions of Suns) accumulates nearby gas resulting in emission from the nucleus, making it a weak Seyfert galaxy or else a LINER ("low-ionization nuclear emission region" – see M 81/NGC 3031) galaxy.

NGC 4559

Constellation	Object type	RA, Dec	Approx. transit date at local midnight	Distance
Coma Berenices	Barred spiral galaxy	12h 36.0 m, +27° 58′	April 24	30 million light years
Age	**Apparent size**	**Magnitude**	**Sky Atlas 2000.0 chart**	**Herald–Bobroff chart**
	11.0′ × 4.9′	10.0	7	C30

This galaxy is near the North Galactic Pole, meaning that there is little foreground extinction of its light from gas and dust in our Galaxy (since its direction lies perpendicular to the disk of our Galaxy). This is also why there are so many distant objects (i.e. galaxies) visible in the constellations surrounding this area. In professional telescopic studies, NGC 4559 contains many ionized hydrogen (HII) regions like M 42. This galaxy has a mass of about 60 billion Suns. Its inclination angle is about 70° (meaning it rotates about an axis that is inclined at 70° from our line-of-sight).

NGC 4565

Constellation	Object type	RA, Dec	Approx. transit date at local midnight	Distance
Coma Berenices	Spiral galaxy	12h 36.3 m, +25° 59'	April 25	30 million light years
Age	**Apparent size**	**Magnitude**	**Sky Atlas 2000.0 chart**	**Herald–Bobroff chart**
	14.9' × 2.0'	9.6	7	C30

With an inclination angle of 87°, this galaxy is virtually edge-on. It has an optical diameter of about 130 thousand light years. This galaxy is quite similar to our Milky Way, although NGC 4565 is slightly more massive (our Galaxy has a mass of about 8 hundred billion Suns). It is thought to have a dark matter halo with a mass of several hundred billion Suns. Its disk contains two ultraluminous X-ray sources that are each thought to be caused by an accretion disk surrounding a small, hundred solar mass black hole. About 2 hundred globular clusters are known in this galaxy from professional telescopic studies. This galaxy is thought to be a Seyfert (see NGC 3227). It is part of a gravitationally bound galaxy triplet with NGC 4562 (barred spiral, 12' SW, mag. 13.5) and NGC 4494 (see NGC 4494, 1° 07' W). With our edge-on view, in professional telescopes its neutral hydrogen (HI) is found to have a twisted ("integral sign") shape, with the twist mostly outside the optically visible portion of the galaxy. Such twists (also called warps, depending on the orientation of our view of the galaxy) occur in the HI distribution of about half of all galactic disks. This galaxy is near the North Galactic Pole, meaning that there is little foreground extinction of its light from gas and dust in our Galaxy (since its direction lies perpendicular to the disk of our Galaxy). This is also why there are so many distant objects (i.e. galaxies) visible in the constellations surrounding this area.

NGC 4567

Constellation	Object type	RA, Dec	Approx. transit date at local midnight	Distance
Virgo	Spiral galaxy	12h 36.5 m, +11° 15'	April 25	55 million light years
Age	**Apparent size**	**Magnitude**	**Sky Atlas 2000.0 chart**	**Herald–Bobroff chart**
	3.1' × 2.2'	11.3	14	C48/D33/E09

Together with nearby NGC 4568 the two galaxies are nicknamed the "Siamese Twins", although it is uncertain whether they are joined or are, more likely, simply the result of an optical illusion, with NGC 4568 actually lying in the foreground. Both belong to the Virgo galaxy cluster (see M 49/NGC 4472).

NGC 4569 (M 90)

Constellation	Object type	RA, Dec	Approx. transit date at local midnight	Distance
Virgo	Barred spiral galaxy	12h 36.8m, +13° 10'	April 26	55 million light years
Age	**Apparent size**	**Magnitude**	**Sky Atlas 2000.0 chart**	**Herald–Bobroff chart**
	9.9' × 4.4'	9.5	14	C48/D33/E09

It is part of the Virgo cluster (see M 49/NGC 4472). It is a LINER galaxy ("low-ionization nuclear emission region" – see M 81/NGC 3031), but its nuclear emission is thought to be dominated by intense star formation (i.e. a starburst) and not by accretion onto a central black hole. About two new stars form every year in this galaxy (about half the number forming yearly in the Milky Way). The galaxy rotates about an axis that is inclined to our line-of-sight by about 64° with the western edge of the galaxy closest to us. Although the galaxy IC 3583 (6' NNW) is nearby in the sky and is also a Virgo cluster member, the two are not thought to be interacting strongly. Because of M 90's high speed relative to the Virgo cluster (over a thousand km/s), it is thought to have lost some of its gas to "ram-pressure stripping" (see M 91/NGC 4548).

NGC 4570

Constellation	Object type	RA, Dec	Approx. transit date at local midnight	Distance
Virgo	Lenticular galaxy	12h 36.9 m, +07° 15'	April 25	55 million light years
Age	**Apparent size**	**Magnitude**	**Sky Atlas 2000.0 chart**	**Herald–Bobroff chart**
	3.7' × 1.2'	10.9	14	C48/D34/E11

It is part of the Virgo galaxy cluster (see M 49/NGC 4472). NGC 4570 has a total luminous mass of about 60 billion Suns and is nearly edge-on. Professional telescopes show it has an outer disk (at radius greater than 7″), two stellar rings (at radii of 1.7″ and 4.5″), and an inner nuclear disk (inside the rings). This complex structure is thought to have been caused by a nuclear bar (which has since faded) that drove gas into the central regions of this galaxy. Some of this material formed into the present-day rings via resonant interactions (where density waves are in resonance with the local speed of epicycle oscillations in the orbits of matter in the disk, including the so-called Lindblad resonances).

NGC 4579 (M 58)

Constellation	Object type	RA, Dec	Approx. transit date at local midnight	Distance
Virgo	Barred spiral galaxy	12h 37.7m, +11° 49'	April 25	55 million light years
Age	**Apparent size**	**Magnitude**	**Sky Atlas 2000.0 chart**	**Herald–Bobroff chart**
	6.0' × 4.8'	9.7	14	C47/D33/E09

It is part of the Virgo galaxy cluster (see M 49/NGC 4472). M 58 has a mass of about 3 hundred million Suns. It is a Seyfert galaxy (or possibly a LINER – see M 81) in which emission from the nucleus is thought to occur due to accretion of matter onto a supermassive central black hole. Its optical diameter is about 1 hundred thousand light years.

NGC 4590 (M 68)

Constellation	Object type	RA, Dec	Approx. transit date at local midnight	Distance
Hydra	Globular cluster	12h 39.5m, –26° 45′	April 26	30 thousand light years
Age	**Apparent size**	**Magnitude**	**Sky Atlas 2000.0 chart**	**Herald–Bobroff chart**
12–14 billion years	9.8′	7.8	21	C66

It has a mass of about 3 hundred thousand Suns. Its size of 9.8′ corresponds to a diameter of almost one hundred light years. Like most globular clusters, it does not orbit our Galaxy with the Galactic disk as we do. Instead it follows an orbit that takes it out as far as about 1 hundred thousand light years from the Galactic center and then back in as close as about 30 thousand light years on an elliptical path (with eccentricity 0.5) inclined to the Galactic disk (by about 30°), taking about 1/2 billion years to make one revolution around our Galaxy. It is very "metal-poor" (meaning it is sparse in elements heavier than helium).

NGC 4594 (M 104)

Constellation	Object type	RA, Dec	Approx. transit date at local midnight	Distance
Virgo	Spiral galaxy	12h 40.0 m, –11° 37′	April 25	30 million light years
Age	**Apparent size**	**Magnitude**	**Sky Atlas 2000.0 chart**	**Herald–Bobroff chart**
	8.6′ × 4.2′	8.0	14	C48/D31

Nicknamed the "Sombrero Galaxy". The well-known dust lane, which actually has a ring-shape (like Saturn's rings), contains about 10 million solar masses worth of dust grains that are submicron in size. The dust lane may owe its existence to the gravitational interaction of a now-defunct bar with the interstellar medium. The disk in this galaxy is about 1/4 as massive as this galaxy's large spheroidal bulge (whereas in our Galaxy the disk has a mass about 7 times that of the bulge). Together the disk and bulge have a mass of several hundred billion Suns. This galaxy harbors a central black hole with a mass of about a billion Suns, onto which a few per cent of a solar mass is accreted every year. It is either a LINER (see M 81/NGC 3031) or Seyfert (see NGC 3227) galaxy, is nearly edge-on (inclination angle of 84°) and contains well over a thousand globular clusters.

NGC 4596

Constellation	Object type	RA, Dec	Approx. transit date at local midnight	Distance
Virgo	Barred lenticular galaxy	12h 39.9 m, +10° 11′	April 25	55 million light years
Age	**Apparent size**	**Magnitude**	**Sky Atlas 2000.0 chart**	**Herald–Bobroff chart**
	4.0′ × 3.4′	10.4	14	C48/D33/E11

It is part of the Virgo galaxy cluster (see M 49/NGC 4472). Its bar takes about 20 million years to make one rotation and accounts for almost half the luminosity of this galaxy. The central bulge in this galaxy accounts for about 30% of its luminosity. The remaining 25% of the luminosity comes from the convex-lens-shaped remaining portion of this galaxy (whence comes the name "lenticular", meaning lens-shaped).

NGC (New General Catalogue) Objects

NGC 4605

Constellation	Object type	RA, Dec	Approx. transit date at local midnight	Distance
Ursa Major	Barred spiral galaxy	12h 40.0 m, +61° 37′	April 25	15 million light years
Age	**Apparent size**	**Magnitude**	**Sky Atlas 2000.0 chart**	**Herald– Bobroff chart**
	5.9′ × 2.4′	10.3	2	C12

This is a dwarf spiral galaxy with an optical diameter of about 25 thousand light years (similar to the diameter of the Large Magellanic Cloud). It is close to edge-on (with an inclination angle of 69°).

NGC 4618

Constellation	Object type	RA, Dec	Approx. transit date at local midnight	Distance
Canes Venatici	Barred Magellanic gal.	12h 41.6 m, +41° 09′	April 26	20 million light years
Age	**Apparent size**	**Magnitude**	**Sky Atlas 2000.0 chart**	**Herald– Bobroff chart**
	4.2′ × 3.4′	10.8	7	C29

This is a classic example of a barred Magellanic galaxy (named after the Large Magellanic Cloud). This galaxy type (which is the "latest" type of spiral galaxy in Hubble's classification scheme before the irregular class) is characterized in professional telescopes by having essentially only one spiral arm, with a bar that is offset from the center of the galaxy. About 1/2 of the light from NGC 4618 comes from its disk, while about 1/4 comes from the one spiral arm and about 1/7 comes from the bar. The remaining 1/10 or so comes from isolated star-forming regions (containing bright, hot young stars and ionized hydrogen – some of these regions are visible as knots in large amateur telescopes). It forms a physical pair with NGC 4625 (8′ NNE, mag. 12.4), which is also a barred Magellanic spiral. The two are thought to be tidally interacting, which is thought to have played a role in causing both galaxies to evolve into the Magellanic spiral classification. The two galaxies have about the same mass, with each at about 50 billion solar masses. Both are part of the M 106 (NGC 4258) galaxy group.

NGC 4621 (M 59)

Constellation	Object type	RA, Dec	Approx. transit date at local midnight	Distance
Virgo	Elliptical galaxy	12h 42.0m, +11° 39′	April 26	55 million light years
Age	**Apparent size**	**Magnitude**	**Sky Atlas 2000.0 chart**	**Herald– Bobroff chart**
	5.3′ × 4.0′	9.6	14	C47/D33/E09

It is part of the Virgo galaxy cluster (see M 49/NGC 4472). The inner core of this galaxy is unusual because it is thought to be counter-rotating from the rest of the galaxy as well as having a different chemical makeup from, and being younger than, the rest of the galaxy, perhaps due to a past accretion event. This inner core has a diameter of several hundred light years and apparent size of about an arcsecond. Professional telescopes find M 59 has a disk that emits about 16% of the light from this galaxy. M 59 also has a circumnuclear disk (diameter of 5–7″) in professional telescopic studies. The center of this galaxy is believed to contain a supermassive black hole (with a mass of about 3 hundred million Suns). M 59 contains about 2 thousand globular clusters.

NGC 4631

Constellation	Object type	RA, Dec	Approx. transit date at local midnight	Distance
Canes Venatici	Barred spiral galaxy	12h 42.1 m, +32° 32′	April 26	25 million light years
Age	Apparent size	Magnitude	Sky Atlas 2000.0 chart	Herald– Bobroff chart
	15.2′ × 2.8′	9.2	7	C29

It is nearly edge-on (with an inclination angle of about 80°) and has an optical diameter of about 110 thousand light years. This galaxy is interacting with its nearby neighbors (NGC 4656 and NGC 4627). The gravitational distortions from these interactions are thought to be triggering intense star formation that is seen throughout the disk of NGC 4631 (as patchy, bright regions in large amateur telescopes). These regions are less than about 10 million years old and together contain about 25 million solar masses. Supernovae and stellar winds from these star-forming regions are thought to be the source of the hot (million degree) corona of gas that surrounds this galaxy. NGC 4631 is part of a group of six gravitationally bound galaxies that includes NGC 4214 (6° 39′ NW – see NGC 4214), NGC 4244 (7° 17′ NW – see NGC 4244), NGC 4395 (3° 34′ WNW), NGC 4656 (33′ SE – see NGC 4656) and UGC 7698 (2° 11′ SW).

NGC 4636

Constellation	Object type	RA, Dec	Approx. transit date at local midnight	Distance
Virgo	Elliptical galaxy	12h 42.8 m, +02° 41′	April 26	55 million light years
Age	Apparent size	Magnitude	Sky Atlas 2000.0 chart	Herald– Bobroff chart
	10.4′ × 5.9′	9.5	14	C47/D34/E11

It is part of the Virgo galaxy cluster (see M 49/NGC 4472), lying at the southern edge of this cluster. This galaxy harbors a supermassive central black hole with a mass of about 80 million Suns (compare this to the 2.6 million solar mass black hole at the center of our Galaxy). Like many such black holes in elliptical galaxies, this one emits unexpectedly little radiation due to accretion, for reasons that remain unclear. One explanation is that the accretion disk is truncated at a certain, inner radius by a jet that extracts energy from the disk and truncates the disk's emission. This galaxy contains nearly 4 thousand globular clusters, which is slightly richer in globulars than most ellipticals in the Virgo cluster.

NGC 4643

Constellation	Object type	RA, Dec	Approx. transit date at local midnight	Distance
Virgo	Barred lenticular galaxy	12h 43.3 m, +01° 59'	April 26	30 million light years
Age	Apparent size	Magnitude	Sky Atlas 2000.0 chart	Herald–Bobroff chart
	3.1' × 2.5'	10.8	14	C47/D34

This is a LINER galaxy (see M 81/NGC 3031). It is the namesake member of the NGC 4643 group of 12 gravitationally bound galaxies that includes the following NGC catalogue entries: 4753 (4° SE – see NGC 4753), 4771 (2° 37' ESE), 4845 (3° 42' E), 4904 (4° 52' ESE), 4900 (4° 21' E), 4713 (3° 44' NNE), 4808 (3° 53' NE).

NGC 4649 (M 60)

Constellation	Object type	RA, Dec	Approx. transit date at local midnight	Distance
Virgo	Elliptical galaxy	12h 43.7m, +11° 33'	April 26	55 million light years
Age	Apparent size	Magnitude	Sky Atlas 2000.0 chart	Herald–Bobroff chart
	7.6' × 6.2'	8.8	14	C47/D33/E09

It is part of the Virgo galaxy cluster (see M 49/NGC 4472). M 60 is thought to contain a supermassive central black hole (with a mass of perhaps a billion Suns). Its nearby companion galaxy NGC 4647 (barred spiral galaxy, 2.5' NW, mag. 11.3) may lie somewhat in front of M 60, but this remains uncertain. The distance between the centers of these two in the sky corresponds to an actual distance of about 40 thousand light years. M 60 contains about 5 thousand globular clusters (about 30 times as many as in our Galaxy).

NGC 4654

Constellation	Object type	RA, Dec	Approx. transit date at local midnight	Distance
Virgo	Barred spiral galaxy	12h 43.9 m, +13° 08'	April 26	55 million light years
Age	Apparent size	Magnitude	Sky Atlas 2000.0 chart	Herald–Bobroff chart
	5.0' × 3.1'	10.5	14	C47/D33/E09

It is part of the Virgo galaxy cluster (see M 49/NGC 4472). Professional telescopic studies find that the neutral hydrogen in this galaxy is distributed asymmetrically, with a large tail off the SE edge that may have been blown out (by so-called "ram pressure") because of this galaxy's Frisbee-like relative motion through the intracluster medium (the material between galaxies in this cluster). In professional telescopes, the northern of its two spiral arms is much stronger, perhaps caused by the off-centeredness of this galaxy's bar. Its optical diameter is about 80 thousand light years.

NGC 4656

Constellation	Object type	RA, Dec	Approx. transit date at local midnight	Distance
Canes Venatici	Irregular galaxy	12h 44.0 m, +32° 10'	April 26	25 million light years
Age	**Apparent size**	**Magnitude**	**Sky Atlas 2000.0 chart**	**Herald–Bobroff chart**
	15.3' × 2.4'	10.5	7	C29

It is nearly edge-on (with an inclination angle of about 80°) and has an optical diameter of about 110 thousand light years. It is part of the NGC 4631 galaxy group (see NGC 4631). Its interaction with nearby NGC 4631 is thought to have triggered star formation in NGC 4656. This starburst activity has blown out chimney-like ionized filaments, the largest of which is a double filament with a mass of several tens of millions of Suns that extends perpendicular to the disk for thousands of light years in professional telescopes. A bridge of neutral hydrogen (with a mass of several billions Suns) extends between NGC 4656 and NGC 4631. NGC 4657 refers to the NE part of NGC 4656.

NGC 4660

Constellation	Object type	RA, Dec	Approx. transit date at local midnight	Distance
Virgo	Elliptical galaxy	12h 44.5 m, +11° 11'	April 27	55 million light years
Age	**Apparent size**	**Magnitude**	**Sky Atlas 2000.0 chart**	**Herald–Bobroff chart**
	2.1' × 1.7'	11.2	14	C47/D33/E09

It contains a supermassive black hole with a mass of several hundred million Suns. It is part of the Virgo galaxy cluster (see M 49/NGC 4472). It is an unusual elliptical galaxy since it contains a disk structure (inclined at about 66° from our line-of-sight) that emits a third of the light from this galaxy.

NGC 4665

Constellation	Object type	RA, Dec	Approx. transit date at local midnight	Distance
Virgo	Barred lenticular galaxy	12h 45.1 m, +03° 03'	April 27	55 million light years
Age	**Apparent size**	**Magnitude**	**Sky Atlas 2000.0 chart**	**Herald–Bobroff chart**
	3.5' × 3.5'	10.5	14	C47/D34/E11

This galaxy is part of the Virgo galaxy cluster (see M 49/NGC 4472). About 20% of all disk galaxies are lopsided (i.e. not azimuthally symmetric), and this is one such galaxy. NGC 4624 and NGC 4664 are repeat entries of NGC 4665.

NGC 4666

Constellation	Object type	RA, Dec	Approx. transit date at local midnight	Distance
Virgo	Barred spiral galaxy	12h 45.1 m, –00° 28′	April 27	85 million light years
Age	**Apparent size**	**Magnitude**	**Sky Atlas 2000.0 chart**	**Herald– Bobroff chart**
	4.5′ × 1.4′	10.7	14	C47/D34

It is nearly edge-on (having an inclination angle of 80°) with the SE edge toward us and the NW edge on the far side of the galaxy. Intense star formation (i.e. a "starburst") is occurring in this galaxy, particularly within the central 1′ radius of its disk, with about one hundred luminous HII (ionized hydrogen) star-forming regions spread over its disk in professional telescopic images. Stellar winds and supernovae from the central star-formation have strewn gas out tens of thousands of light years perpendicular to the disk in a cone-shaped "superwind" that is thought to have been ionized mainly by shock waves. Less intense star formation farther out in the disk of the galaxy is still intense enough that it is thought to have blown gas into an extended gaseous halo that surrounds this galaxy. Tidal disturbances caused by nearby NGC 4668 (7′ SE) may be responsible for triggering the starburst in NGC 4666. These two galaxies along with NGC 4632 (45′ WNW) form a gravitationally bound triplet.

NGC 4689

Constellation	Object type	RA, Dec	Approx. transit date at local midnight	Distance
Coma Berenices	Spiral galaxy	12h 47.8 m, +13° 46′	April 27	55 million light years
Age	**Apparent size**	**Magnitude**	**Sky Atlas 2000.0 chart**	**Herald– Bobroff chart**
	4.7′ × 4.0′	10.9	14	C47/D33/E09

It is part of the Virgo galaxy cluster (see M 49/NGC 4472), lying at its eastern edge. As with many of the spirals in the Virgo cluster, much of the gas from this galaxy has been stripped away by interaction with the intracluster medium (see NGC 4419).

NGC 4697

Constellation	Object type	RA, Dec	Approx. transit date at local midnight	Distance
Virgo	Elliptical galaxy	12h 48.6 m, –05° 48′	April 28	40 million light years
Age	**Apparent size**	**Magnitude**	**Sky Atlas 2000.0 chart**	**Herald– Bobroff chart**
	6.2′ × 4.5′	9.2	14	C47/D31

It has a total mass of about 2 hundred billion Suns, with a central black hole mass of about a hundred million Suns. Professional telescopic studies find more than 5 hundred planetary nebulae in this galaxy (about the same number as in our Galaxy). It is part of the NGC 4697 group of 17 gravitationally bound galaxies that includes nearby NGC 4731 (51′ SE), NGC 4775 (1.5° ESE), NGC 4941 (4° E), NGC 4951 (4° E), NGC 4948 (4.5° ESE), and NGC 4958 (4.8° ESE).

NGC 4698

Constellation	Object type	RA, Dec	Approx. transit date at local midnight	Distance
Virgo	Spiral galaxy	12h 48.4 m, +08° 29'	April 28	55 million light years
Age	**Apparent size**	**Magnitude**	**Sky Atlas 2000.0 chart**	**Herald–Bobroff chart**
	4.0' × 2.9'	10.6	14	C47/D33/E11 (unlabelled)

It is part of the Virgo galaxy cluster (see M 49/NGC 4472), lying at the northeastern boundary of this cluster. This is a Seyfert galaxy (see NGC 3227). Its central bulge rotates about an axis that is perpendicular to the axis of rotation of its disk, the only spiral galaxy known to do so, and thought to be the result of accretion in its past.

NGC 4699

Constellation	Object type	RA, Dec	Approx. transit date at local midnight	Distance
Virgo	Barred spiral galaxy	12h 49.0 m, –08° 40'	April 28	60 million light years
Age	**Apparent size**	**Magnitude**	**Sky Atlas 2000.0 chart**	**Herald–Bobroff chart**
	3.8' × 2.8'	9.5	14	C47/D31

It has an optical diameter of about 65 thousand light years and an inclination angle of about 40°. It is the namesake member of the NGC 4699 galaxy group of 14 gravitationally bound galaxies that includes NGC 4802 (3° 46' SSE), NGC 4722 (4° 41' S), NGC 4700 (2° 45' S), NGC 4742 (2° SSE), NGC 4781 (2° 18' SE), and NGC 4790 (2° SE).

NGC 4725

Constellation	Object type	RA, Dec	Approx. transit date at local midnight	Distance
Coma Berenices	Barred spiral galaxy	12h 50.4 m, +25° 30'	April 28	40 million light years
Age	**Apparent size**	**Magnitude**	**Sky Atlas 2000.0 chart**	**Herald–Bobroff chart**
	10.4' × 7.2'	9.4	7	C29

This galaxy is near the North Galactic Pole, meaning that there is little foreground extinction of its light from gas and dust in our Galaxy (since its direction lies perpendicular to the disk of our Galaxy). This is also why there are so many distant objects (i.e. galaxies) visible in the constellations surrounding this area. This galaxy has a mass of a hundred billion Suns within a radius of 3' of its center. Its optical diameter is 120 thousand light years and rotates about an axis inclined at about 50° from our line-of-sight. It has an inner ring with a diameter of 4.5' in professional telescopes (not to be confused with the incomplete pseudo-ring of its spiral arms at a diameter of 10' that is visible in large amateur telescopes). The inner ring is thought to be expanding at about 50 km/s and rotating at just over 2 hundred km/s, and is undergoing active star formation. This galaxy is interacting with nearby NGC 4747 (25' NE), the latter being about 1/20 the mass of NGC 4725. The galaxy NGC 4712 (11' W) is a background object.

NGC 4736 (M 94)

Constellation	Object type	RA, Dec	Approx. transit date at local midnight	Distance
Canes Venatici	Spiral galaxy	12h 50.9 m, +41° 07′	April 29	15 million light years

Age	Apparent size	Magnitude	Sky Atlas 2000.0 chart	Herald–Bobroff chart
	12.3′ × 10.8′	8.2	7	C29/D30

It has a mass of about 60 billion Suns and an optical diameter of about 50 thousand light years. Professional telescopes find M 94 has an inner ring (45″ in radius), as well as an outer ring (5.5′ in radius). The inner ring is undergoing intense star formation (i.e. it is a "starburst" ring), creating about 1.5 new stars/year. The rings are thought to be the result of orbital resonant interactions where density waves are in resonance with the local speed of epicycle oscillations in the orbits of matter in the disk, including the so-called Lindblad resonances. Most of the galaxy rotates at approximately 150 km/s about an axis inclined to our line-of-sight by about 40°. M 94 is a LINER galaxy (see M 81/NGC 3031) and is part of the M 106 group of 17 gravitationally bound galaxies (see M 106/NGC 4258).

NGC 4753

Constellation	Object type	RA, Dec	Approx. transit date at local midnight	Distance
Virgo	Lenticular galaxy	12h 52.4 m, –01° 12′	April 28	30 million light years

Age	Apparent size	Magnitude	Sky Atlas 2000.0 chart	Herald–Bobroff chart
	5.1′ × 2.6′	9.9	14	C47/D34

Professional telescopic studies find this galaxy has a complicated set of dust lanes that are thought to be the result of a nearly edge-on view of this galaxy's "twisted disk" (meaning the plane of the disk varies with radius). The twisted disk is thought to be due to the accretion of a gas-rich dwarf galaxy that occurred at least a half billion years ago. It is part of the NGC 4643 galaxy group (see NGC 4643).

NGC 4754

Constellation	Object type	RA, Dec	Approx. transit date at local midnight	Distance
Virgo	Barred lenticular galaxy	12h 52.3 m, +11° 19′	April 28	55 million light years

Age	Apparent size	Magnitude	Sky Atlas 2000.0 chart	Herald–Bobroff chart
	4.4′ × 2.4′	10.6	14	C47/D33

It is part of the Virgo galaxy cluster (see M 49/NGC 4472), as are nearby NGC 4733 and NGC 4762 (see NGC 4762). These three galaxies lie at the eastern boundary of the Virgo cluster.

NGC 4762

Constellation	Object type	RA, Dec	Approx. transit date at local midnight	Distance
Virgo	Barred lenticular galaxy	12h 52.9 m, +11° 14'	April 29	55 million light years
Age	**Apparent size**	**Magnitude**	**Sky Atlas 2000.0 chart**	**Herald–Bobroff chart**
	8.6' × 2.0'	10.3	14	C47/D33

It is part of the Virgo galaxy cluster (see M 49/NGC 4472), lying at its eastern edge along with NGC 4733 and NGC 4754. It contains a nuclear disk of gas (3″ in radius) that is rotating more rapidly than the stars there, at a speed of over one hundred km/s at its outer edge. This galaxy's mass out to a radius of 1' from the center is about 35 billion Suns. The galaxy is perfectly edge-on. It is a LINER galaxy (see M 81/NGC 3031). Like the majority of barred lenticular galaxies, an outer ring is seen (at a diameter of 8.3') in professional telescopes.

NGC 4781

Constellation	Object type	RA, Dec	Approx. transit date at local midnight	Distance
Virgo	Barred spiral galaxy	12h 54.4 m, −10° 32'	April 29	60 million light years
Age	**Apparent size**	**Magnitude**	**Sky Atlas 2000.0 chart**	**Herald–Bobroff chart**
	3.4' × 1.4'	11.1	14	C47/D31

It is part of the NGC 4699 galaxy group (see NGC 4699), to which nearby NGC 4742 and NGC 4790 also belong. NGC 4760 (19' W) is a background object and several times farther away.

NGC 4800

Constellation	Object type	RA, Dec	Approx. transit date at local midnight	Distance
Canes Venatici	Spiral galaxy	12h 54.6 m, +46° 32'	April 29	50 million light years
Age	**Apparent size**	**Magnitude**	**Sky Atlas 2000.0 chart**	**Herald–Bobroff chart**
	1.6' × 1.2'	11.5	7	C29

It rotates about an axis that is inclined at about 40° from our line-of-sight. It has an optical diameter of about 20 thousand light years.

NGC 4826 (M 64)

Constellation	Object type	RA, Dec	Approx. transit date at local midnight	Distance
Coma Berenices	Spiral galaxy	12h 56.7 m, +21° 41′	April 30	16 million light years
Age	Apparent size	Magnitude	Sky Atlas 2000.0 chart	Herald–Bobroff chart
	10.3′ × 5.0′	8.5	7	C29

Nicknamed the "Black-Eye Galaxy" (because of a dark arc-shaped dust region on its NE side, which is a challenge to discern in amateur telescopes). It has an optical diameter of about 50 thousand light years. The nucleus of this galaxy is chemically different from the rest of the galaxy (and is said to be "chemically decoupled"). In addition, the gas in the outer disk (radii > about 1′ and containing a hundred million solar masses) counter-rotates from rest of the galaxy, including the stars (which all rotate the same way). This highly unusual situation may have its origin in the past accretion of a gas-rich dwarf satellite galaxy. The dust that gives the galaxy its "black eye" rotates with the stars. M 64 is thought to be a LINER galaxy driven by a starburst in its nucleus (see M 81/NGC 3031 for explanation).

NGC 4845

Constellation	Object type	RA, Dec	Approx. transit date at local midnight	Distance
Virgo	Spiral galaxy	12h 58.0 m, +01° 35′	April 29	30 million light years
Age	Apparent size	Magnitude	Sky Atlas 2000.0 chart	Herald–Bobroff chart
	4.9′ × 1.2′	11.2	14	C47/D34

It is nearly edge-on (with an inclination angle of 78°). Its central bulge is thought to rotate about an axis that is not aligned with either the major or minor axes of its elliptical shape (so the bulge is said to be "triaxial"). It has a number of HII (ionized hydrogen) star-forming regions and is part of the NGC 4643 galaxy group (see NGC 4643).

NGC 4856

Constellation	Object type	RA, Dec	Approx. transit date at local midnight	Distance
Virgo	Barred lenticular galaxy	12h 59.4 m, −15° 03′	April 30	70 million light years
Age	Apparent size	Magnitude	Sky Atlas 2000.0 chart	Herald–Bobroff chart
	3.9′ × 1.4′	10.5	14	C47/D31

It is nearly edge-on (having an inclination angle of about 70°) with an optical diameter of about 80 thousand light years and a position angle of 35°.

NGC 4866

Constellation	Object type	RA, Dec	Approx. transit date at local midnight	Distance
Virgo	Lenticular galaxy	12h 59.5 m, +14° 10′	April 30	50 million light years
Age	**Apparent size**	**Magnitude**	**Sky Atlas 2000.0 chart**	**Herald– Bobroff chart**
	6.4′ × 1.5′	11.2	14	C47/D33

This is a LINER galaxy (see M 81/NGC 3031).

NGC 4900

Constellation	Object type	RA, Dec	Approx. transit date at local midnight	Distance
Virgo	Barred spiral galaxy	13h 00.7 m, +02° 30′	May 1	30 million light years
Age	**Apparent size**	**Magnitude**	**Sky Atlas 2000.0 chart**	**Herald– Bobroff chart**
	2.2′ × 2.2′	11.4	14	C47/D31

It is part of the NGC 4643 galaxy group (see NGC 4643). It is nearly face-on (with an inclination angle of about 5°).

NGC 4958

Constellation	Object type	RA, Dec	Approx. transit date at local midnight	Distance
Virgo	Barred lenticular galaxy	13h 05.8 m, –08° 01′	May 2	40 million light years
Age	**Apparent size**	**Magnitude**	**Sky Atlas 2000.0 chart**	**Herald– Bobroff chart**
	3.9′ × 1.4′	10.7	14	C47/D31

It is part of the NGC 4697 galaxy group (see NGC 4697), as is its near neighbor NGC 4948. It has an optical diameter of about 50 thousand light years.

NGC 4995

Constellation	Object type	RA, Dec	Approx. transit date at local midnight	Distance
Virgo	Barred spiral galaxy	13h 09.7 m, –07° 50′	May 3	90 million light years
Age	**Apparent size**	**Magnitude**	**Sky Atlas 2000.0 chart**	**Herald– Bobroff chart**
	2.4′ × 1.7′	11.1	14	C47/D31

This is a Seyfert galaxy (see NGC 3227). It has many HII (ionized hydrogen, star-forming) regions in its disk. It is the namesake member of the NGC 4995 group of five gravitationally bound galaxies the brightest of which are NGC 4928 (1° 41′ W) and NGC 4981 (1° NNW). Nearby NGC 4958 and NGC 4928 are part of the much closer NGC 4697 galaxy group (see NGC 4697).

NGC (New General Catalogue) Objects

NGC 5005

Constellation	Object type	RA, Dec	Approx. transit date at local midnight	Distance
Canes Venatici	Barred spiral galaxy	13h 10.9 m, +37° 03′	May 3	70 million light years
Age	**Apparent size**	**Magnitude**	**Sky Atlas 2000.0 chart**	**Herald–Bobroff chart**
	5.8′ × 2.9′	9.8	7	C29/D30

It has a mass of several hundred billion Suns. It is alternately classed as a LINER galaxy (see M 81/NGC 3031) or a Seyfert galaxy (see NGC 3227). It is the namesake member of the NGC 5005 group of 13 galaxies that includes nearby NGC 5033 (40′ SE – see NGC 5033), NGC 5002 (26′ S), NGC 5014 (47′ S), IC 4213 (1° 24′ S), NGC 4861 (3° 15′, SW), NGC 5107 (2° 33′ NE), NGC 5112 (2° 45′ NE), IC 4182 (1° 09′ WNW).

NGC 5024 (M 53)

Constellation	Object type	RA, Dec	Approx. transit date at local midnight	Distance
Coma Berenices	Globular cluster	13h 12.9m, +18° 10′	May 4	60 thousand light years
Age	**Apparent size**	**Magnitude**	**Sky Atlas 2000.0 chart**	**Herald–Bobroff chart**
12–14 billion years	14.4′	7.6	14	C29

Its mass is about 3/4 million Suns (which is about a millionth the mass of our entire Galaxy). It lies nearly directly "below" us from the Galactic central plane in the halo of our Galaxy (see M 2/NGC 7089 for the meaning of "halo"). Its orbit keeps it out in the halo, with a period of close to a billion years. It follows a path that takes it well over a hundred thousand light years from the Galactic center and never closer than about 50 thousand light years to the Galactic center. Its orbit is highly inclined (by about 60°) to the disk of our Galaxy. M 53's apparent size corresponds to a diameter of about 250 light years. The nearby globular cluster NGC 5053 (1° ESE) lies nearly directly "above" M 53 in three-dimensional space, being a little less than 10 thousand light years closer to us (in a direction nearly perpendicular to the disk of our Galaxy).

NGC 5033

Constellation	Object type	RA, Dec	Approx. transit date at local midnight	Distance
Canes Venatici	Spiral galaxy	13h 13.5 m, +36° 36′	May 4	60 million light years
Age	**Apparent size**	**Magnitude**	**Sky Atlas 2000.0 chart**	**Herald–Bobroff chart**
	9.8′ × 3.6′	10.2	7	C29/D30

It has a total dynamical mass of about 1/4 trillion Suns. It is thought to have been perturbed in the past by weak interaction with other galaxies, perhaps dwarf galaxies that lie nearby (as the Milky Way has been perturbed by the Magellanic Clouds). It is part of the NGC 5005 galaxy group (see NGC 5005). It is a Seyfert galaxy (see NGC 3227).

NGC 5054

Constellation	Object type	RA, Dec	Approx. transit date at local midnight	Distance
Virgo	Spiral galaxy	13h 17.0 m, –16° 38′	May 5	90 million light years
Age	**Apparent size**	**Magnitude**	**Sky Atlas 2000.0 chart**	**Herald–Bobroff chart**
	5.1′ × 2.8′	10.9	14	C47/D31

It has a mass of about 140 billion Suns and an optical diameter of 130 thousand light years. It is thought to be a member of the NGC 5044 galaxy group that covers roughly two square degrees centered on NGC 5044 and whose NGC members are NGC 5017, 5030, 5031, 5035, 5037, 5038, 5044, 5046, 5047, 5049 and 5054 itself, but which includes over one hundred much dimmer galaxies that lie in this region as well.

NGC 5055 (M 63)

Constellation	Object type	RA, Dec	Approx. transit date at local midnight	Distance
Canes Venatici	Spiral galaxy	13h 15.8 m, +42° 02′	May 4	30 million light years
Age	**Apparent size**	**Magnitude**	**Sky Atlas 2000.0 chart**	**Herald–Bobroff chart**
	12.6′ × 7.5′	8.6	7	C29/D30

Nicknamed the "Sunflower Galaxy". It is part of a gravitationally bound group of four galaxies that includes M 51 (NGC 5194) and NGC 5023 (spiral galaxy, mag. 12.3, 2° NNW), in addition to the dimmer UGC 8320 (irregular galaxy, mag. 13, 4° N). M 63 is a so-called "flocculent" spiral galaxy, meaning that it lacks any obvious azimuthally symmetric spiral arm pattern (see NGC 3521 for further explanation). It has a mass of about 140 billion Suns and is a "low-ionization nuclear emission region" galaxy (or "LINER" – see M 81/NGC 3031 for explanation). It has an optical diameter of a little over 1 hundred thousand light years. Beyond its optically visible regions the galaxy consists of a neutral hydrogen (HI) disk (not visible in amateur telescopes) that is "warped" so that the far outer edge of its HI disk lies in a plane that is skewed by perhaps about 20° from the rest of the galaxy. Warps occur in the HI distribution of about half of all galactic disks.

NGC 5194 (M 51)

Constellation	Object type	RA, Dec	Approx. transit date at local midnight	Distance
Canes Venatici	Spiral galaxy	13h 29.9m, +47° 12′	May 8	30 million light years
Age	**Apparent size**	**Magnitude**	**Sky Atlas 2000.0 chart**	**Herald–Bobroff chart**
	10.8′ × 6.6′	8.4	7	C29

Nicknamed the "Whirlpool Galaxy". It is a grand-design spiral, meaning that it has two symmetrically placed spiral arms that extend over most its visible disk in professional telescopic images. What appears as a bridge connecting M 51 to its nearby companion galaxy NGC 5195 (4′ NNE – see NGC 5195) is actually an optical illusion – a spiral arm of M 51 is superimposed on NGC 5195 with NGC 5195 actually lying on the far side of M 51 (by perhaps 1/2 million light years), although the two galaxies have had one or more close encounters in the past (see NGC 5195). Gravitational interaction with NGC 5195 is thought to be triggering star formation in M 51, with about four new stars per year forming in M 51. This is similar to the number forming per year in the Milky Way, but note that our Galaxy has a mass about 10 times that of M 51. More than a thousand ionized hydrogen (HII) star-forming regions have been identified in M 51 by the Hubble Space Telescope, with diameters averaging about a hundred light years and having masses up to about a thousand Suns. A number of these HII regions can be seen as bright knots in amateur telescopes. M 51 is a LINER galaxy (see M 81/NGC 3031 for explanation) and has an optical diameter of almost 1 hundred thousand light years.

NGC 5195

Constellation	Object type	RA, Dec	Approx. transit date at local midnight	Distance
Canes Venatici	Barred lenticular galaxy	13h 30.0 m, +47° 16′	May 8	30 million light years
Age	**Apparent size**	**Magnitude**	**Sky Atlas 2000.0 chart**	**Herald–Bobroff chart**
	5.9′ × 4.6′	9.6	7	C29

This is the companion galaxy with which M 51 is interacting (see M 51 / NGC 5194). This interaction is thought to be responsible for triggering star formation in the arms of M 51, and has had a strong effect on the location and strength of M 51's star formation. NGC 5195 is on the far side of M 51 from us (by perhaps 1/2 million light years) and thought to be several times less massive than M 51. Some models suggest that NGC 5195 is orbiting M 51 (with a period of about a hundred million of years, with the last closest approach about 70 million years ago), but its orbit is decaying so that it may merge with M 51 (perhaps in less than a few hundred million years). Other models suggest the two had a single "hit-and-run" collision much longer ago – perhaps nearly a billion years ago. NGC 5195 is thought to have had a nuclear starburst in its past, which has since ceased but whose evolved stars still yield considerable emission from the nucleus of this galaxy. Indeed, it is a LINER galaxy (see M 81/NGC 3031). It has an optical diameter of about 50 thousand light years.

NGC 5236 (M 83)

Constellation	Object type	RA, Dec	Approx. transit date at local midnight	Distance
Hydra	Barred spiral galaxy	13h 37.0m, −29° 52′	May 11	12 million light years
Age	**Apparent size**	**Magnitude**	**Sky Atlas 2000.0 chart**	**Herald–Bobroff chart**
	13.1′ × 12.2′	7.5	21	C65/D29

This is a starburst galaxy (meaning it has intense star formation occurring – see M 82/NGC 3034) with the star formation concentrated in a half-ringlet occupying the region 3–7″ from the galaxy center that contains hundreds of star clusters. About thirty of these star clusters have masses of more than 20 thousands Suns and are less than 10 million years old. M 83 is part of a gravitationally bound group of galaxies that includes nearby NGC 5264 (1° E) and NGC 5253 (1° 53′ SSE), the latter having its closest approach to M 83 one or 2 billion years ago. M83 is currently in second place in the contest for the galaxy with the most recorded supernovae, having 6, which is second only to the 7 recorded in NGC 6946. In common with M 31, professional telescopes find M 83 may have a double nucleus (each containing about 130 million solar masses and separated by 2.7″ at a position angle of 243° i.e. aligned along a WSW–ENE direction). M 83 has an optical diameter of about 50 thousand light years and is nearly face-on (with an inclination angle of 24° i.e. it rotates about an axis that is inclined from our line-of-sight by 24°).

NGC 5248

Constellation	Object type	RA, Dec	Approx. transit date at local midnight	Distance
Bootes	Barred spiral galaxy	13h 37.5 m, +08° 53′	May 10	70 million light years
Age	**Apparent size**	**Magnitude**	**Sky Atlas 2000.0 chart**	**Herald–Bobroff chart**
	5.9′ × 4.5′	10.3	14	C47

This is a grand-design spiral (meaning that it has two symmetrically placed spiral arms that extend over most of its visible disk in professional telescopic images). It also has a nuclear grand-design spiral (3″ in radius) that is coupled to the outer spiral arms, so that the spiral pattern of this galaxy covers an amazing range of length scales from hundreds of light years to several tens of thousands of light years. Professional telescopic studies find a 14″ diameter circumnuclear ring of "super star clusters" (each containing tens of thousands of stars) that may be globular clusters at a young age (a few tens of million years). NGC 5248 has a mass of about 140 billion Suns and an optical diameter of about 120 thousand light years. The NE side of this galaxy is the side closest to us. It is a Seyfert galaxy (see NGC 3227).

NGC 5272 (M 3)

Constellation	Object type	RA, Dec	Approx. transit date at local midnight	Distance
Canes Venatici	Globular cluster	13h 42.2m, +28° 23′	May 11	30 thousand light years
Age	**Apparent size**	**Magnitude**	**Sky Atlas 2000.0 chart**	**Herald– Bobroff chart**
12–14 billion years	18.6′	6.2	7	C29

Its mass is nearly 8 hundred thousand Suns. Its size of 18.6′ corresponds to a diameter of about 160 light years. It orbits the Galaxy on a precessing elliptical path that is highly inclined with the plane of our Galaxy and quite eccentric (minor axis to major axis ratio of 0.4). It takes about 3 hundred million years to make one revolution of the Galaxy, never straying farther than about 50 thousand light years from the Galactic center (we are about 25 thousand light years from the Galactic center, although we stay in the disk of our Galaxy and M 3 does not). M 3 never approaches closer than about 15 thousand light years from the Galactic center, making it an inner halo cluster (so called since it doesn't travel too far out in the halo – see M 2 for the meaning of "halo").

NGC 5273

Constellation	Object type	RA, Dec	Approx. transit date at local midnight	Distance
Canes Venatici	Lenticular galaxy	13h 42.1 m, +35° 39′	May 11	70 million light years
Age	**Apparent size**	**Magnitude**	**Sky Atlas 2000.0 chart**	**Herald– Bobroff chart**
	2.7′ × 2.2′	11.6	7	C29/D30

This is a Seyfert galaxy (where a supermassive object in this galaxy's center accumulates nearby material resulting in strong emission from the nucleus). It has an optical diameter of about 60 thousand light years, which is about the average diameter of nearby galaxies (i.e. those within 150 million light years).

NGC 5322

Constellation	Object type	RA, Dec	Approx. transit date at local midnight	Distance
Ursa Major	Elliptical galaxy	13h 49.2 m, +60° 11′	May 13	1 hundred million light years
Age	**Apparent size**	**Magnitude**	**Sky Atlas 2000.0 chart**	**Herald– Bobroff chart**
	6.0′ × 4.1′	10.2	2	C11

This is a LINER galaxy (see M 81/NGC 3031). It has a central core within a radius of 10″ that counter-rotates relative to the rest of the galaxy, which is thought to be the result of a merger of two galaxies several billion years ago. It is the namesake member of the NGC 5322 galaxy group, which contains eight gravitationally bound galaxies with the following NGC catalogue members: 5322, 5308, 5379, 5389, 5376, 5372, 5342.

NGC 5363

Constellation	Object type	RA, Dec	Approx. transit date at local midnight	Distance
Virgo	Lenticular galaxy	13h 56.1 m, +05° 15'	May 15	70 million light years
Age	**Apparent size**	**Magnitude**	**Sky Atlas 2000.0 chart**	**Herald–Bobroff chart**
	4.6' × 3.1'	10.1	14	C47

Professional telescopic studies find a neutral hydrogen (HI) disk surrounding this galaxy that is falling into the galaxy at several tens of km/s, resulting in accretion of a few percent of a solar mass/year into the inner regions of the galaxy. It is a LINER ("low-ionization nuclear emission region") galaxy (see M 81/NGC 3031) and is part of the NGC 5364 galaxy group (see NGC 5364).

NGC 5364

Constellation	Object type	RA, Dec	Approx. transit date at local midnight	Distance
Virgo	Spiral galaxy	13h 56.2 m, +05° 01'	May 15	80 million light years
Age	**Apparent size**	**Magnitude**	**Sky Atlas 2000.0 chart**	**Herald–Bobroff chart**
	6.1' × 4.2'	10.5	14	C47

Professional telescopes find this galaxy has an amazingly regular spiral appearance – its arms can be traced for well over a complete revolution from their origin. Significant HII (ionized hydrogen) star-forming regions are present. It is the namesake member of the NGC 5364 galaxy group of seven gravitationally bound galaxies that includes nearby NGC 5356, NGC 5363 (14' N – see NGC 5363), NGC 5338 and NGC 5300. One degree north is the NGC 5374 galaxy group, which is more than twice as far away as NGC 5364 and includes NGC 5374 and 5382.

NGC 5457 (M 101)

Constellation	Object type	RA, Dec	Approx. transit date at local midnight	Distance
Ursa Major	Barred spiral galaxy	14h 03.2 m, +54° 21'	May 17	25 million light years
Age	**Apparent size**	**Magnitude**	**Sky Atlas 2000.0 chart**	**Herald–Bobroff chart**
	28.5' × 28.3'	7.9	2	C11

This galaxy is nearly face-on. It has a diameter of almost 2 hundred thousand light years. Many ionized hydrogen (HII) star-forming regions are present in M 101, some of which can be seen as bright knots in amateur telescopes. Several of these knots have their own NGC numbers e.g. NGC 5461, 5462, 5471, the latter of which is a hundred times larger and brighter than any HII region in our Galaxy. Some of these supersized HII regions have masses of tens of millions of Suns (e.g. NGC 5461 and 5471), and although they appear as a single knot in amateur telescopes, they are made up of many individual giant molecular clouds (GMCs) that have masses of several hundred thousands of Suns (similar to the masses of GMCs in our Galaxy). M 101 is the namesake member of the M 101 group of gravitationally bound galaxies that includes NGC 5474 (44' SSE), NGC 5585 (3° 21' NE), NGC 5204 (6° NW) and NGC 5477 (22' ENE) and Holmberg IV (UGC 8837, 1° 19' WSW). Tidal interactions with several group members in the past few hundred million years (perhaps including NGC 5477, NGC 5474 and Holmberg IV) are thought to have distorted M 101. These interactions may have induced the formation of some of the HII regions e.g. NGC 5471.

NGC 5466

Constellation	Object type	RA, Dec	Approx. transit date at local midnight	Distance
Bootes	Globular cluster	14h 05.5 m, +28° 32′	May 17	55 thousand light years
Age	Apparent size	Magnitude	Sky Atlas 2000.0 chart	Herald– Bobroff chart
12–14 billion years	9.2′	9.0	7	C28

It orbits the Galaxy with a period of about a billion years, never getting any closer than about 20 thousand light years to the Galactic center, but wandering out over 2 hundred thousand light years at its most distant orbital position. This cluster has an orbit that shares dynamical properties with that of NGC 6934 (see NGC 6934), and it has been suggested that both are left over from a disrupted satellite galaxy of the Milky Way. NGC 5466 has a diameter of about 150 light years and a mass of about 130 thousand Suns. Although binary stars are rare in globular clusters, professional telescopes find this cluster contains three eclipsing binaries (two of which are "contact binaries", meaning they are so close to each other that they are immersed in a common envelope of gas).

NGC 5473

Constellation	Object type	RA, Dec	Approx. transit date at local midnight	Distance
Ursa Major	Barred lenticular galaxy	14h 04.7 m, +54° 54′	May 17	110 million light years
Age	Apparent size	Magnitude	Sky Atlas 2000.0 chart	Herald– Bobroff chart
	2.2′ × 1.7′	11.4	2	C10

It is part of the NGC 5485 galaxy group of seven gravitationally bound galaxies that includes nearby NGC 5422, NGC 5443, NGC 5475, NGC 5485 and NGC 5486. Nearby M 101 (35′ SSW) is a foreground object, being more than six times closer.

NGC 5474

Constellation	Object type	RA, Dec	Approx. transit date at local midnight	Distance
Ursa Major	Spiral galaxy	14h 05.0 m, +53° 40′	May 17	20 million light years
Age	Apparent size	Magnitude	Sky Atlas 2000.0 chart	Herald– Bobroff chart
	4.5′ × 3.3′	10.8	2	C10

It has a mass of several billion Suns. It is nearly face-on (with an inclination angle of about 20°). Professional telescopes find that the disk of this galaxy is warped at its outer edges (in its neutral hydrogen disk beyond the optically visible region), becoming increasingly less face-on at larger radii (reaching inclination angles of 30–40°). Such warps occur in the neutral hydrogen (HI) distribution of about half of all galactic disks. This galaxy is interacting with its much larger neighbor, M 101, which is thought to have strongly disturbed it. It is part of the M 101 galaxy group (see M 101/NGC 5457).

NGC 5557

Constellation	Object type	RA, Dec	Approx. transit date at local midnight	Distance
Bootes	Elliptical galaxy	14h 18.4 m, +36° 30′	May 21	150 million light years
Age	**Apparent size**	**Magnitude**	**Sky Atlas 2000.0 chart**	**Herald–Bobroff chart**
	2.4′ × 1.9′	11.0	7	C28

It forms a gravitationally bound galaxy trio with the pair NGC 5590 (1° 29′ SSE) and NGC 5589 (1° 24′ SSE).

NGC 5566

Constellation	Object type	RA, Dec	Approx. transit date at local midnight	Distance
Virgo	Barred spiral galaxy	14h 20.3 m, +03° 56′	May 21	90 million light years
Age	**Apparent size**	**Magnitude**	**Sky Atlas 2000.0 chart**	**Herald–Bobroff chart**
	6.6′ × 2.3′	10.6	14	C46

This is a LINER galaxy (LINER stands for "low-ionization nuclear emission region" – see M 81/NGC 3031) and is thought to contain two bars. It is inclined at about 80° (i.e. nearly edge-on). It is the namesake member of the NGC 5566 group of six gravitationally bound galaxies that includes nearby NGC 5560 (5′ NW), 5574 (43′ SSE), 5576 (42′ SSE – see NGC 5576) and 5577 (33′ SSE) as well as the dimmer NGC 5569 (4′ NE).

NGC 5576

Constellation	Object type	RA, Dec	Approx. transit date at local midnight	Distance
Virgo	Elliptical galaxy	14h 21.1 m, +03° 16′	May 21	90 million light years
Age	**Apparent size**	**Magnitude**	**Sky Atlas 2000.0 chart**	**Herald–Bobroff chart**
	3.0′ × 2.3′	11.0	14	C46

It is part of the NGC 5566 galaxy group (see NGC 5566). The nucleus of this galaxy is much younger (being only a few billion years old) and has a different chemical makeup compared to the rest of the galaxy. In addition, the nucleus rotates in a plane perpendicular to the main plane of the galaxy. This decoupled nucleus is thought to be the result of a past accretion of material with its own, separate, chemical makeup and angular velocity.

NGC 5631

Constellation	Object type	RA, Dec	Approx. transit date at local midnight	Distance
Ursa Major	Lenticular galaxy	14h 26.6 m, +56° 35′	May 23	110 million light years
Age	**Apparent size**	**Magnitude**	**Sky Atlas 2000.0 chart**	**Herald–Bobroff chart**
	2.0′ × 2.0′	11.5	2	C10

This is a Seyfert galaxy (where a supermassive object in this galaxy's center accumulates nearby material resulting in strong emission from the nucleus).

NGC 5634

Constellation	Object type	RA, Dec	Approx. transit date at local midnight	Distance
Virgo	Globular cluster	14h 29.6 m, −05° 59′	May 23	90 thousand light years
Age	**Apparent size**	**Magnitude**	**Sky Atlas 2000.0 chart**	**Herald–Bobroff chart**
12–14 billion years	3.7′	9.5	14	C46

It has a mass of nearly 3 hundred thousand Suns and a diameter of about 90 light years. It lies well off the Galactic plane (by about 60 thousand light years). It has been suggested that it once belonged to the Sagittarius Dwarf Elliptical Galaxy (see M 54/NGC 6715), but was stripped away by gravitational interactions with our Galaxy billions of years ago.

NGC 5676

Constellation	Object type	RA, Dec	Approx. transit date at local midnight	Distance
Bootes	Spiral galaxy	14h 32.8m, +49° 27′	May 24	110 million light years
Age	**Apparent size**	**Magnitude**	**Sky Atlas 2000.0 chart**	**Herald–Bobroff chart**
	3.9′ × 1.8′	11.2	7	C10

It has a mass of about 2 hundred billion Suns. It is the namesake member of the NGC 5676 group of galaxies that includes 11 gravitationally bound galaxies, the brightest of which include NGC 5660, NGC 5689 (see NGC 5689), NGC 5673, IC 1029 (all within 1° of NGC 5676), as well as NGC 5707 (2° NNE) and NGC 5633 (3.5° SSW).

NGC 5689

Constellation	Object type	RA, Dec	Approx. transit date at local midnight	Distance
Bootes	Barred lenticular galaxy	14h 35.5m, +48° 44′	May 26	110 million light years
Age	**Apparent size**	**Magnitude**	**Sky Atlas 2000.0 chart**	**Herald–Bobroff chart**
	3.3′ × 1.0′	11.9	7	C10

It is part of the NGC 5676 group of galaxies (see NGC 5676) that also includes the much dimmer NGC 5682 (mag. 14.1), and NGC 5693 (mag. 13.5).

NGC 5694

Constellation	Object type	RA, Dec	Approx. transit date at local midnight	Distance
Hydra	Globular cluster	14h 39.6m, −26° 32′	May 26	110 thousand light years
Age	**Apparent size**	**Magnitude**	**Sky Atlas 2000.0 chart**	**Herald–Bobroff chart**
12–14 billion years	2.2′	10.2	21	C64

It has a mass of about 3 hundred thousand Suns and a diameter of about 70 light years. It has an unusually low abundance of elements heavier than helium (i.e. it is "metal-poor"), even for a globular cluster, belying its old age. It lies in the halo (see NGC 7006) of our Galaxy, one of only 13 outer halo globular clusters (that lie more than 80 thousand light years from the Galactic center).

NGC 5746

Constellation	Object type	RA, Dec	Approx. transit date at local midnight	Distance
Virgo	Barred spiral galaxy	14h 44.9m, +01° 57′	May 27	90 million light years
Age	**Apparent size**	**Magnitude**	**Sky Atlas 2000.0 chart**	**Herald–Bobroff chart**
	6.9′ × 1.2′	10.3	14	C46

It has an optical diameter of about 180 thousand light years and is nearly edge-on (with an inclination angle of 82°). It is part of the NGC 5668 galaxy group of 22 gravitationally bound galaxies, among which are the following nearby readily visible NGC objects: 5690, 5740, 5746, 5692, 5691, 5705, 5713, 5719, 5750, 5638, 5701, 5668, as well as IC 1024 and IC 1048.

NGC 5846

Constellation	Object type	RA, Dec	Approx. transit date at local midnight	Distance
Virgo	Elliptical galaxy	15h 06.5m, +01° 36′	June 2	90 million light years
Age	**Apparent size**	**Magnitude**	**Sky Atlas 2000.0 chart**	**Herald–Bobroff chart**
	4.0′ × 3.7′	10.0	14	C46

This is a large galaxy with a mass of several trillion Suns. It is the namesake member of the NGC 5846 group of galaxies that all lie at approximately the same distance, and which includes all the readily visible NGC galaxies within a 3° × 3° square, most of which are "early-type" galaxies (i.e. ellipticals, lenticulars and early-type spirals). Dust found in the central region of NGC 5846 in professional telescopic studies is thought to be from a past accretion of a small, gas-rich galaxy. A high-speed encounter with NGC 5850 is thought to have left the latter with a disturbed morphology.

NGC 5866 (M 102)

Constellation	Object type	RA, Dec	Approx. transit date at local midnight	Distance
Draco	Lenticular galaxy	15h 06.5m, +55° 46′	June 2	50 million light years
Age	**Apparent size**	**Magnitude**	**Sky Atlas 2000.0 chart**	**Herald–Bobroff chart**
	6.5′ × 3.1′	9.9	2	C09/C10

This is a lenticular galaxy (given the label S0 in classification schemes), meaning it has a disk and central bulge like a spiral galaxy, but lacks the spiral arms. Lenticulars contain very little gas, dust, or young stars, consisting almost entirely of old stars. This galaxy is sometimes called the "Spindle Galaxy", although NGC 3115 also has this nickname. It is nearly edge-on (inclination angle of 71°, meaning it rotates about an axis that is inclined at 71° from our line-of-sight) with about 60% of its light coming from its bulge and 40% from its disk. It has an optical diameter of 90 thousand light years (the diameter of our Galaxy is about 1 hundred thousand light years). It is the namesake member of the NGC 5866 group of galaxies that includes nearby NGC 5907 and NGC 5879, which are within several million light years of NGC 5866. Messier's original M 102 is believed to be a duplicate entry of M 101 (NGC 5457) rather than being NGC 5866, but in order to have 110 different objects in the Messier list, many amateur astronomers informally attach the label M 102 to NGC 5866.

NGC 5897

Constellation	Object type	RA, Dec	Approx. transit date at local midnight	Distance
Libra	Globular cluster	15h 17.4m, −21° 01′	June 4	40 thousand light years
Age	**Apparent size**	**Magnitude**	**Sky Atlas 2000.0 chart**	**Herald–Bobroff chart**
12–14 billion years	8.7′	8.5	14	C63

It has a mass of about 2 hundred thousand Suns and a diameter of about one hundred light years. Like 3/4 of the globular clusters in our Galaxy, it is "metal-poor" (meaning it contains relatively low amounts of elements heavier than helium). It contains about 40 "blue stragglers", which are stars that are paradoxically far more blue and luminous than expected for reasons that remain unknown but may involve the coalescence of stars (see NGC 6633).

NGC 5904 (M 5)

Constellation	Object type	RA, Dec	Approx. transit date at local midnight	Distance
Serpens Caput	Globular cluster	15h 18.6m, +02° 05′	June 5	24 thousand light years
Age	**Apparent size**	**Magnitude**	**Sky Atlas 2000.0 chart**	**Herald–Bobroff chart**
12–14 billion years	19.9′	5.7	15	C45

Its mass is over 8 hundred thousand Suns and its diameter is about 140 light years. Like most globular clusters it does not orbit our Galaxy with the Galactic disk as we do. Instead it follows an orbit that takes it out as far as 150 thousand light years from the Galactic center and then back in as close as about 10 thousand light years, on a path highly inclined to the Galactic disk, taking almost a billion years to make one revolution around our Galaxy (compare to the 1/4 billion years our Sun takes to make one Galactic revolution). M 5 currently sits about 20 thousand light years from the Galactic center, which is much closer than its average orbital distance.

NGC 5907 (NGC 5906)

Constellation	Object type	RA, Dec	Approx. transit date at local midnight	Distance
Draco	Spiral galaxy	15h 15.9m, +56° 20′	June 5	45 million light years
Age	**Apparent size**	**Magnitude**	**Sky Atlas 2000.0 chart**	**Herald–Bobroff chart**
	11.8′ × 1.3′	10.3	2	C10

Its nearly edge-on size of 11.8′ gives it an optical diameter of about 150 thousand light years. Although not visible in amateur telescopes, it has a faint ring which is thought to be the remains of a tidally disrupted companion dwarf spheroidal galaxy (that had a mass of 2 hundred million Suns and was disrupted by NGC 5907 about a billion years ago). It also has an extended halo which may be due to cannibalization of a small elliptical galaxy. It is part of the NGC 5866 galaxy group (see M 102/NGC 5866).

NGC 5982

Constellation	Object type	RA, Dec	Approx. transit date at local midnight	Distance
Draco	Elliptical galaxy	15h 38.7m, +59° 21′	June 11	1 hundred million light years
Age	**Apparent size**	**Magnitude**	**Sky Atlas 2000.0 chart**	**Herald–Bobroff chart**
	3.0′ × 2.1′	11.1	2	C10

As with about 25% of elliptical galaxies, the core of this galaxy rotates independently of its outer regions, a feature which may result from the merging of two galaxies. In such mergers, the central part retains the angular momentum of one of the merged galaxies, while the outer parts retain the angular momentum of the other galaxy. It is part of the NGC 5985 group of four gravitationally bound galaxies that includes NGC 5985 (7′ E, a Seyfert galaxy – see NGC 3227), as well as NGC 5987 (1° 17′ S) and NGC 5989 (32′ NE). Nearby NGC 5981 (6′ WNW) is not considered part of this group but lies at a similar distance.

NGC 6093 (M 80)

Constellation	Object type	RA, Dec	Approx. transit date at local midnight	Distance
Scorpius	Globular cluster	16h 17.0m, −22° 59′	June 21	30 thousand light years
Age	**Apparent size**	**Magnitude**	**Sky Atlas 2000.0 chart**	**Herald–Bobroff chart**
12–14 billion years	5.1′	7.3	22	C63/D13

It has a mass of nearly 4 hundred thousand Suns and a diameter of about 50 light years. It was the first globular cluster to have a nova discovered in it. M 80 is a bulge cluster (meaning it orbits inside the central bulge of our Galaxy – see M 9/NGC 6333) and has one of the shortest orbital periods of the globular clusters in our Galaxy. Indeed, it only takes about 70 million years to complete one revolution about the Galaxy, in an orbit that is highly inclined to the Galactic central plane.

NGC 6118

Constellation	Object type	RA, Dec	Approx. transit date at local midnight	Distance
Serpens Caput	Spiral galaxy	16h 21.8m, –02° 17′	June 21	80 million light years
Age	**Apparent size**	**Magnitude**	**Sky Atlas 2000.0 chart**	**Herald– Bobroff chart**
	4.7′ × 1.9′	11.7	15	C45

It has a mass of about one hundred billion Suns, an optical diameter of about 1 hundred billion light years, and an inclination angle of about 60° (meaning its disk rotates about an axis that is inclined at about 60° from our line-of-sight).

NGC 6121 (M 4)

Constellation	Object type	RA, Dec	Approx. transit date at local midnight	Distance
Scorpius	Globular cluster	16h 23.6m, –26° 32′	June 21	7 thousand light years
Age	**Apparent size**	**Magnitude**	**Sky Atlas 2000.0 chart**	**Herald– Bobroff chart**
12–14 billion years	22.8′	5.6	22	C63/D13

This is the nearest globular cluster to us. Its size of 22.8′ corresponds to a diameter of about 50 light years. It has a mass of about 2 hundred thousand Suns. It is the only globular cluster known to harbor a triple star system. Even more unusual is that this is the only known triple star system that contains a pulsar (see M 1 for explanation of pulsar). This triple star consists of a neutron star with a mass about 1.4 times that of our Sun, a white dwarf with a mass about 30% that of our Sun, and a "planet" with < 1% the mass of our Sun. This planet may have been stolen from another star–planet system by the neutron star in the past few million years. The cluster lies toward the Galactic center, within roughly two thousand light years of the Galactic central plane, so that interstellar material in the disk of our Galaxy blocks out some of its light and makes it dimmer (by a few magnitudes) than it would otherwise appear.

NGC 6144

Constellation	Object type	RA, Dec	Approx. transit date at local midnight	Distance
Scorpius	Globular cluster	16h 27.2m, –26° 01′	June 22	30 thousand light years
Age	**Apparent size**	**Magnitude**	**Sky Atlas 2000.0 chart**	**Herald– Bobroff chart**
12–14 billion years	6.2′	9.0	22	C63/D13

It has a mass of about 150 thousand Suns. Its size of 6.2′ gives it a diameter of about 50 light years. It lies behind the ρ Ophiuchi dust cloud which, along with other inter-stellar material between us and the cluster, dims our view of the stars in this cluster. It is a bulge cluster (i.e. it orbits in the central ball-like bulge of our Galaxy) and is metal-poor (i.e. it contains low amounts of elements heavier than helium), so that it may have formed in the earliest star-forming period of our Galaxy (see NGC 6287).

NGC 6171 (M 107)

Constellation	Object type	RA, Dec	Approx. transit date at local midnight	Distance
Ophiuchus	Globular cluster	16h 32.5m, –13° 03′	June 23	20 thousand light years

Age	Apparent size	Magnitude	Sky Atlas 2000.0 chart	Herald–Bobroff chart
12–14 billion years	3.3′	8.1	15	C45

It has a mass of about 2 hundred thousand Suns and a diameter of about 20 light years. It is a bulge cluster (i.e. it orbits in the central ball-like bulge of our Galaxy) with a period of about one hundred million years. Its orbital path is inclined by about 45° to the Galactic disk and is a very flattened ellipse. It was not noted by Messier, but instead is a recent (1947) addition to the Messier catalogue suggested by H.S. Hogg.

NGC 6205 (M 13)

Constellation	Object type	RA, Dec	Approx. transit date at local midnight	Distance
Hercules	Globular cluster	16h 41.7m, +36° 28′	June 26	25 thousand light years

Age	Apparent size	Magnitude	Sky Atlas 2000.0 chart	Herald–Bobroff chart
12–14 billion years	17′	5.8	8	C26

Nicknamed the "Hercules Cluster". Its 17′ size corresponds to a diameter of about 120 light years. It contains about 6 hundred thousand solar masses. It orbits the Galaxy independently of the material in the Galactic disk on an inclined orbit that travels out to about 80 thousand light years from the Galactic center but approaches within 20 thousand light years of the Galactic center, taking a 1/2 billion or so years to complete one revolution.

NGC 6207

Constellation	Object type	RA, Dec	Approx. transit date at local midnight	Distance
Hercules	Spiral galaxy	16h 43.1m, +36° 50′	June 26	55 million light years

Age	Apparent size	Magnitude	Sky Atlas 2000.0 chart	Herald–Bobroff chart
	3.0′ × 1.2′	11.6	8	C26

It has an optical diameter of about 50 thousand light years and a mass of about 15 billion Suns. A foreground star 6″ N of the nucleus is partly responsible for the star-like appearance of the nucleus. It rotates about an axis inclined at 64° from our line-of-sight.

NGC 6210

Constellation	Object type	RA, Dec	Approx. transit date at local midnight	Distance
Hercules	Planetary nebula	16h 44.5m, +23° 48′	June 27	4 thousand light years
Age	**Apparent size**	**Magnitude**	**Sky Atlas 2000.0 chart**	**Herald–Bobroff chart**
	14″	9	8	C26

The three-dimensional structure of this nebula is complex and not well understood, with professional telescopic studies finding extensions in several directions that are thought to be jets ejected from the nucleus, in addition to expanding shell structures. The central star (mag. 13.7) has a surface temperature of about 70 thousand K and can be seen in large amateur telescopes.

NGC 6217

Constellation	Object type	RA, Dec	Approx. transit date at local midnight	Distance
Ursa Minor	Barred spiral galaxy	16h 32.6m, +78° 12′	June 24	80 million light years
Age	**Apparent size**	**Magnitude**	**Sky Atlas 2000.0 chart**	**Herald–Bobroff chart**
	3.3′ × 3.2′	11.2	3	C09

This is a Seyfert galaxy (see NGC 3227) with a nuclear starburst (i.e. intense nuclear star formation). It has one of the largest (apparent size) one-sided X-ray jets, extending 2.7′ SW from its nucleus in professional telescopic studies, presumably associated with accretion onto a supermassive central object.

NGC 6218 (M 12)

Constellation	Object type	RA, Dec	Approx. transit date at local midnight	Distance
Ophiuchus	Globular cluster	16h 47.2m, −01° 57′	June 27	22 thousand light years
Age	**Apparent size**	**Magnitude**	**Sky Atlas 2000.0 chart**	**Herald–Bobroff chart**
12–14 billion years	12.2′	6.7	15	C44

Its mass is about 1/4 million Suns. Its size of 12.2′ corresponds to a diameter of about 80 light years. Like M 3 and M 10, this is an inner halo cluster, so called since it doesn't travel too far out in the halo – see M 2 for the meaning of "halo". M 12 never strays farther than about 20 thousand light years from the Galactic center on an orbit inclined to the Galactic central plane by 33° or so. M 12 takes about 130 million years to complete one revolution around the Galaxy, having just crossed the Galactic central plane a few million years ago (lying 2 thousand light years below it) on its way to a maximum excursion of a little under 10 thousand light years below the Galactic plane. (Contrast this with the Sun, which stays in the 2 thousand light year thick disk of the Galaxy and is currently about 50 light years above the Galactic central plane, having crossed the central plane about 2 million years ago, on its way to a maximum excursion of about 250 light years above the Galactic central plane.)

NGC 6229

Constellation	Object type	RA, Dec	Approx. transit date at local midnight	Distance
Hercules	Globular cluster	16h 47.0m, +47° 32′	June 27	1 hundred thousand light years
Age	**Apparent size**	**Magnitude**	**Sky Atlas 2000.0 chart**	**Herald–Bobroff chart**
12–14 billion years	3.8′	9.4	8	C09

It has a mass of about half a million Suns. Its size of 3.8′ gives it a diameter of about 110 light years. It lies in the outer halo of our Galaxy (see NGC 7006 for explanation of "halo"), 65 thousand light years above the Galactic central plane and about 1 hundred thousand light years from the Galactic center. Such "outer halo" globular clusters are uncommon – only 13 of the 150 or so globulars in our Galaxy are farther than 80 thousand light years from the Galactic center, with the only other outer halo NGC globular clusters being NGC 2419 (see NGC 2419), NGC 5694 (see NGC 5694), NGC 5824, and NGC 7006 (see NGC 7006).

NGC 6235

Constellation	Object type	RA, Dec	Approx. transit date at local midnight	Distance
Ophiuchus	Globular cluster	16h 53.4m, −22° 11′	June 29	30 thousand light years
Age	**Apparent size**	**Magnitude**	**Sky Atlas 2000.0 chart**	**Herald–Bobroff chart**
12–14 billion years	1.9′	10.0	22	C62/D13

It has a mass of nearly 2 hundred Suns. It has a diameter of almost 20 light years. It lies on the other side of the Galactic center from us (10 thousand light years from the Galactic center), about 7 thousand light years below the Galactic central plane. It is a bulge cluster (i.e. it lies in the central ball-like bulge of our Galaxy) and is metal-poor (i.e. it contains low amounts of elements heavier than helium), so that it may have formed in an early star-forming period of our Galaxy (see NGC 6287).

NGC 6254 (M 10)

Constellation	Object type	RA, Dec	Approx. transit date at local midnight	Distance
Ophiuchus	Globular cluster	16h 57.1m, −04° 06′	June 30	15 thousand light years
Age	**Apparent size**	**Magnitude**	**Sky Atlas 2000.0 chart**	**Herald–Bobroff chart**
	12.2′	6.6	15	C44

Its mass is a little over 2 hundred thousand Suns. Its size of 12.2′ corresponds to a diameter of about 50 light years. This cluster stays within several thousand light years of the Galactic central plane and has a velocity close to that of the material in the Galactic disk. This is quite unusual for a "metal-poor" cluster like this one ("metal-poor" meaning it has low amounts of elements heavier than helium), since stars in the Galactic disk tend to have "metal" that was scattered by previous supernovae. The origin of this cluster is thus uncertain. Models predict this cluster will undergo core collapse in perhaps about 10 billion years, a process that causes stars near the cluster center to confine themselves to an inordinately small region of space like a swarm of angry bees suddenly placed into a small container – see NGC 6284. Core collapse will be followed by destruction of the cluster by gravitational interactions in about twice this time.

NGC 6266 (M 62)

Constellation	Object type	RA, Dec	Approx. transit date at local midnight	Distance
Ophiuchus	Globular cluster	17h 01.2m, –30° 07′	July 1	20 thousand light years
Age	**Apparent size**	**Magnitude**	**Sky Atlas 2000.0 chart**	**Herald–Bobroff chart**
12–14 billion years	14.1′	6.5	22	C62/D13

It has a mass of almost a million Suns. Its size of 14.1′ corresponds to a diameter of about 80 light years. It lies on the edge of the Galactic disk, in the direction of the Galactic center from us and is a "bulge cluster" (meaning it spends its time orbiting around the central, spherical, 15 thousand light year diameter bulge of our Galaxy). It is one of fewer than 20 globular clusters known to contain a pulsar (with a pulse period of 5.24 milliseconds; see M 1 for the meaning of "pulsar"). The pulsar in this case is a binary star consisting of a neutron star and a white dwarf (orbiting the neutron star every 3.8 days). The core of M 62 is thought to be "collapsed" (see NGC 6284 for explanation).

NGC 6273 (M 19)

Constellation	Object type	RA, Dec	Approx. transit date at local midnight	Distance
Ophiuchus	Globular cluster	17h 02.6m, –26° 16′	July 1	30 thousand light years
Age	**Apparent size**	**Magnitude**	**Sky Atlas 2000.0 chart**	**Herald–Bobroff chart**
12–14 billion years	12.3′	6.2	22	C62/D13

Its mass is about 1.5 million Suns. This is the most flattened globular cluster known in our Galaxy (its oval shape is apparent in amateur telescopes). It gives off the second most light of all the NGC globular clusters in our Galaxy, after ω Centauri (NGC 5139). It belongs to the central bulge of our Galaxy (i.e. the central, spherically shaped region), lying a few thousand light years nearly directly above the Galactic center. It is a metal-poor bulge cluster (metal-poor meaning it has a low abundance of elements heavier than helium) and so may have formed in the earliest star-forming period of our Galaxy (see NGC 6287).

NGC 6284

Constellation	Object type	RA, Dec	Approx. transit date at local midnight	Distance
Ophiuchus	Globular cluster	17h 04.5m, –24° 46′	July 1	50 thousand light years
Age	**Apparent size**	**Magnitude**	**Sky Atlas 2000.0 chart**	**Herald–Bobroff chart**
12–14 billion years	2.7′	8.8	22	C62/D13

This object has a mass of about 2 hundred thousand Suns. Its size of 2.7′ gives it a diameter of about 40 light years. It lies well on the other side of the Galactic center from us (20 thousand light years from the Galactic center), about 8 thousand light years below the Galactic central plane. The core of this cluster has "collapsed", a process that results from an instability that causes stars near the cluster center to confine themselves to an inordinately small region of space (with interstellar spacings as low as the distance between the Sun and Pluto), like a swarm of angry bees suddenly placed into a small container. About 20% of globular clusters in our Galaxy have collapsed cores.

NGC 6287

Constellation	Object type	RA, Dec	Approx. transit date at local midnight	Distance
Ophiuchus	Globular cluster	17h 05.2m, −22° 42′	July 2	30 thousand light years
Age	**Apparent size**	**Magnitude**	**Sky Atlas 2000.0 chart**	**Herald–Bobroff chart**
12–14 billion years	2.7′	9.4	22	C62/D13

It has a mass of about 1 hundred thousand Suns. This is a very metal-poor cluster (meaning it contains very low amounts of elements heavier than helium i.e. "metals"). It lies in the central bulge of our Galaxy. It has been proposed that such metal-poor bulge globular clusters formed early in the Galaxy's life, during a burst of star-forming activity from primordial matter (that lacks metals) that ended with the ejection of gas by these new stars and their supernovae. Subsequent in-fall of this gas is thought to have led to later bursts of star forming (but now with more metals, since some of the gas had already been converted to metals in the first set of stars), with the later bursts giving birth to the more metal-rich globular clusters in the bulge. Having possibly participated in the first star-forming burst, NGC 6287 is thought to be one of the oldest recognizable single objects in the Galaxy.

NGC 6293

Constellation	Object type	RA, Dec	Approx. transit date at local midnight	Distance
Ophiuchus	Globular cluster	17h 10.2m, −26° 35′	July 3	30 thousand light years
Age	**Apparent size**	**Magnitude**	**Sky Atlas 2000.0 chart**	**Herald–Bobroff chart**
12–14 billion years	3.5′	8.2	22	C62/D13

It has a mass of about 2 hundred thousand Suns and a diameter of about 30 light years. It lies in the central bulge of our Galaxy and is a very metal-poor cluster (i.e. it contains very low amounts of elements heavier than helium), so that it may have formed in the earliest star-forming period of our Galaxy (see NGC 6287). This is a core-collapsed cluster (see NGC 6284 for explanation).

NGC 6304

Constellation	Object type	RA, Dec	Approx. transit date at local midnight	Distance
Ophiuchus	Globular cluster	17h 14.5m, −29° 28′	July 3	20 thousand light years
Age	**Apparent size**	**Magnitude**	**Sky Atlas 2000.0 chart**	**Herald–Bobroff chart**
12–14 billion years	3.8′	8.2	22	C62/D13

It lies on this side of the Galactic center, close to the Galactic central plane. It has a mass of about 2 hundred thousand Suns. It is a bulge cluster and is "metal-rich" (meaning it has significant amounts of elements heavier than helium) and so may have formed somewhat later than nearby metal-poor bulge clusters (see NGC 6287). It lies in a window of low foreground extinction, like Baade's window (see NGC 6522), that allows us to peer farther through space here than in other nearby areas of the sky because of reduced amounts of interstellar material.

NGC 6316

Constellation	Object type	RA, Dec	Approx. transit date at local midnight	Distance
Ophiuchus	Globular galaxy	17h 16.6m, −28° 08′	July 4	40 thousand light years
Age	**Apparent size**	**Magnitude**	**Sky Atlas 2000.0 chart**	**Herald– Bobroff chart**
12–14 billion years	2.4′	8.4	22	C62/D13

It has a mass of about a million Suns and a diameter of about 30 light years. It lies in the central bulge of our Galaxy, on the other side of the Galactic center from us, a few thousand light years below the Galactic central plane. It is a "metal-rich" cluster (meaning it has significant amounts of elements heavier than helium) and so may have formed somewhat later than nearby metal-poor bulge clusters (see NGC 6287).

NGC 6333 (M 9)

Constellation	Object type	RA, Dec	Approx. transit date at local midnight	Distance
Ophiuchus	Globular cluster	17h 19.2m, −18° 31′	July 5	27 thousand light years
Age	**Apparent size**	**Magnitude**	**Sky Atlas 2000.0 chart**	**Herald– Bobroff chart**
12–14 billion years	5.5′	7.7	15	C62/D13

Its mass is about 3 hundred thousand Suns. Its size of 5.5′ gives it a diameter of about 40 thousand light years. It belongs to the central bulge of our Galaxy (i.e. the central, spherically shaped region within about 15 thousand light years of the Galactic center) and lies a few thousand light years nearly directly below the Galactic center. This is a metal-poor bulge cluster (metal-poor meaning it has a low abundance of elements heavier than helium) and so may have formed in the earliest star-forming period of our Galaxy (see NGC 6287).

NGC 6341 (M 92)

Constellation	Object type	RA, Dec	Approx. transit date at local midnight	Distance
Hercules	Globular cluster	17h 17.1m, +43° 08′	July 6	27 thousand light years
Age	**Apparent size**	**Magnitude**	**Sky Atlas 2000.0 chart**	**Herald– Bobroff chart**
12–14 billion years	12.2′	6.4	8	C26

This is one of the oldest globular clusters known and is very metal-poor (i.e. having a low abundance of elements heavier than helium). It has a mass of about 4 hundred thousand Suns. Its size of 12.2′ corresponds to a diameter of about one hundred light years. It orbits the Galaxy on a path that is somewhat inclined to the disk (by a little more than 20°) that takes it a maximum of about 13 thousand light years away from the Galactic central plane, traveling between about 5 and 35 thousand light years from the Galactic center. It completes one revolution around the Galaxy about once every 2 hundred million years.

NGC 6342

Constellation	Object type	RA, Dec	Approx. transit date at local midnight	Distance
Ophiuchus	Globular cluster	17h 21.2m, −19° 35′	July 6	30 thousand light years
Age	**Apparent size**	**Magnitude**	**Sky Atlas 2000.0 chart**	**Herald– Bobroff chart**
12–14 billion years	1.3′	9.7	15	C62/D13

It has a mass of about 2 hundred thousand Suns and a diameter of about 10 light years. It lies in the central bulge of our Galaxy. It contains an eclipsing binary pulsar (i.e. a rapidly rotating neutron star with an eclipsing companion). The presence of this pulsar is puzzling since they result from supernovae and this cluster shouldn't have had any supernovae for many billions of years. Thus, any pulsars that were in this cluster should have stopped spinning long ago. One explanation is that an old pulsar was spun up again by a recent (< 10 million year old) interaction with another star. Another explanation is that a white dwarf in the cluster recently (within 10 million years) accreted matter from a companion and went supernova, thereby forming the pulsar. In either case, the pulsar is part of a binary system whose companion has been detected in professional telescopic studies. The companion has a mass of only a few tenths of our Sun, and is very faint at mag. 25. This is a core-collapsed cluster (see NGC 6284 for explanation).

NGC 6355

Constellation	Object type	RA, Dec	Approx. transit date at local midnight	Distance
Ophiuchus	Globular cluster	17h 24.0m, −26° 21′	July 6	20 thousand light years
Age	**Apparent size**	**Magnitude**	**Sky Atlas 2000.0 chart**	**Herald– Bobroff chart**
12–14 billion years	6.1′	9.1	22	C62/D13

It has a mass of almost 4 hundred thousand Suns and a diameter of about 40 light years. It is a bulge cluster (i.e. it lies in the central ball-like bulge of our Galaxy) and very metal-poor (i.e. it contains very low amounts of elements heavier than helium), so that it may have formed in the earliest star-forming period of our Galaxy (see NGC 6287). This is a core-collapsed cluster (see NGC 6284 for explanation).

NGC 6356

Constellation	Object type	RA, Dec	Approx. transit date at local midnight	Distance
Ophiuchus	Globular cluster	17h 23.6m, –17° 49'	July 6	50 thousand light years
Age	Apparent size	Magnitude	Sky Atlas 2000.0 chart	Herald–Bobroff chart
12–14 billion years	3.5'	8.3	15	C62/D13

It has a mass of about 8 hundred thousand Suns and its size of 3.5' gives it a diameter of about 50 light years. It lies on the other side of the Galactic center from us in the central bulge of our Galaxy. It is a "metal-rich" cluster (meaning it has significant amounts of elements heavier than helium) and so may have formed somewhat later than nearby metal-poor bulge clusters (see NGC 6287). It completes an orbit of the Galaxy every few hundred million years, having crossed the Galactic central plane a few tens of millions of years ago.

NGC 6369

Constellation	Object type	RA, Dec	Approx. transit date at local midnight	Distance
Ophiuchus	Planetary nebula	17h 29.3m, –23° 46'	July 8	6 thousand light years
Age	Apparent size	Magnitude	Sky Atlas 2000.0 chart	Herald–Bobroff chart
	30"	11	22	C62/D13

Nicknamed the "Little Ghost Nebula". It is expanding at about 40 km/s. Its central star (mag. 15.6) is a Wolf–Rayet star (see NGC 40 for explanation) with a mass of about 65% that of our Sun and a temperature of about 70 thousand K.

NGC 6401

Constellation	Object type	RA, Dec	Approx. transit date at local midnight	Distance
Ophiuchus	Globular cluster	17h 38.6m, –23° 55'	July 10	40 thousand light years
Age	Apparent size	Magnitude	Sky Atlas 2000.0 chart	Herald–Bobroff chart
12–14 billion years	1.0'	9.5	22	C62/D13

It contains over a million Suns and its size of 1.0' gives it a diameter of about 10 light years. It is a bulge cluster (i.e. it lies in the central ball-like bulge of our Galaxy), lying on the far side of the bulge from us. It is moderately metal-poor (i.e. it contains low amounts of elements heavier than helium), so that it may have formed in an early star-forming period of our Galaxy (see NGC 6287).

NGC 6402 (M 14)

Constellation	Object type	RA, Dec	Approx. transit date at local midnight	Distance
Ophiuchus	Globular cluster	17h 37.6m, –03° 15′	July 10	30 thousand light years

Age	Apparent size	Magnitude	Sky Atlas 2000.0 chart	Herald–Bobroff chart
12–14 billion years	6.7′	7.6	15	C44

Its mass is about 1.2 million times that of our Sun. Its size of 6.7′ corresponds to a diameter of about 60 light years. It is located in the central bulge of our Galaxy (i.e. the central, spherically shaped region) and is relatively lacking in elements heavier than helium (i.e. "metals"), so that it may have formed in one of the earliest star-forming periods of our Galaxy (see NGC 6287). It lies about 8 thousand light years above the Galactic central plane about 13 thousand light years from the Galactic center. It was only the second globular cluster (after M 80) to have a nova discovered in it.

NGC 6405 (M 6)

Constellation	Object type	RA, Dec	Approx. transit date at local midnight	Distance
Scorpius	Open cluster	17h 40.3m, –32° 15′	July 10	1.6 thousand light years

Age	Apparent size	Magnitude	Sky Atlas 2000.0 chart	Herald–Bobroff chart
80 million years	20′	4.0	22	C62/D12

Nicknamed the "Butterfly Cluster" due to the shape of its apparent outline. It contains several chemically peculiar stars ("CP2" stars – see NGC 7243 and NGC 2169 for explanation). It lies in the direction of the center of our Galaxy, less than twenty light years below the Galactic central plane. Its size of 20′ corresponds to a diameter of about 10 light years. Professional telescopic studies have counted over 3 hundred stars belonging to this cluster (only a small fraction of which are visible in amateur telescopes).

NGC 6426

Constellation	Object type	RA, Dec	Approx. transit date at local midnight	Distance
Ophiuchus	Globular cluster	17h 44.9m, +03° 10′	July 12	60 thousand light years

Age	Apparent size	Magnitude	Sky Atlas 2000.0 chart	Herald–Bobroff chart
12–14 billion years	2.2′	11.0	15	C44

It lies 16 thousand light years from the Galactic plane, and 37 thousand light years from the Galactic center. Its size of 2.2′ gives it a size of 10 light years. It has a mass of nearly a hundred thousand Suns. It is one of the most "metal-poor" globulars in our Galaxy (meaning it has lower than usual amounts of elements heavier than helium).

NGC 6440

Constellation	Object type	RA, Dec	Approx. transit date at local midnight	Distance
Sagittarius	Globular cluster	17h 48.9m, −20° 22′	July 14	30 thousand light years
Age	Apparent size	Magnitude	Sky Atlas 2000.0 chart	Herald–Bobroff chart
12–14 billion years	1.7′	9.2	15	C62/D11

This is one of 20 or so of the most massive globular clusters in our Galaxy, containing nearly 6 hundred thousand solar masses. It is one of only a dozen or so globular clusters that is a bright X-ray source, thought to be caused by interacting binary stars (i.e. a neutron star and a low-mass companion). Its size of 1.7′ gives it a diameter of about 15 light years. It is a bulge cluster (meaning it lies in the central bulge of our Galaxy) and is "metal-rich" (meaning it has significant amounts of elements heavier than helium), so that it may have formed somewhat later than nearby metal-poor bulge clusters (see NGC 6287).

NGC 6445

Constellation	Object type	RA, Dec	Approx. transit date at local midnight	Distance
Sagittarius	Planetary nebula	17h 49.3m, −20° 01′	July 14	7 thousand light years
Age	Apparent size	Magnitude	Sky Atlas 2000.0 chart	Herald–Bobroff chart
	34″	13	15	C62/D11

The nebula is believed to have evolved from a progenitor star with a mass > 3 solar masses. It is one of only about 14% of planetary nebulae that are not classified as round or elliptical. It is instead bipolar, meaning it has a three-dimensional "bowtie" shape in professional telescopes. It was discovered in 1882 by Pickering and is about 1 light year in diameter.

NGC 6451

Constellation	Object type	RA, Dec	Approx. transit date at local midnight	Distance
Scorpius	Open cluster	17h 50.7m, −30° 13′	July 15	7 thousand light years
Age	Apparent size	Magnitude	Sky Atlas 2000.0 chart	Herald–Bobroff chart
2 hundred million years	7′	8.0	22	C61

An apparent diameter of 7′ gives it an actual size of 14 light years. It was discovered in 1784 by William Herschel.

NGC 6475 (M 7)

Constellation	Object type	RA, Dec	Approx. transit date at local midnight	Distance
Scorpius	Open cluster	17h 53.9m, –34° 48′	July 14	1 thousand light years
Age	Apparent size	Magnitude	Sky Atlas 2000.0 chart	Herald–Bobroff chart
2 hundred million years	80′	3.3	22	C62/D12

First mentioned by Ptolemy over 2 thousand years ago, it has a diameter of about 20 light years. Professional telescopic studies find that nearly 100% of the stars in M 7 are binary stars, an inordinately high frequency of binaries compared to the Galactic field (where > 50% of main-sequence stars are binaries). Professional telescopic studies have counted about a hundred stars belonging to this cluster. Like all open clusters, a good number of the stars in the field of this cluster (called "field stars") do not belong to the cluster. Instead, they just happen to lie in the line-of-sight of the cluster but are actually much closer or farther away from us than the cluster. This "contamination" by field stars can be seen in Figure 6475 (see M 39 for further discussion of this).

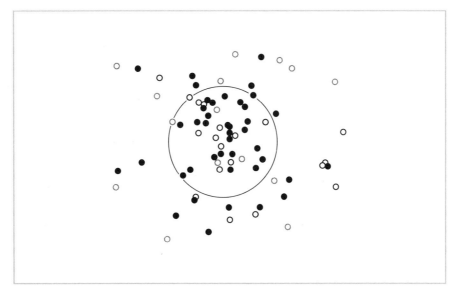

Figure 6475 The brighter stars (< mag. 10) in the region of M 7 are shown. The large circle is shown to give a length scale and has a diameter of 60′. Stars shown by a solid circle can be considered as belonging to M 7 since they have a greater than 90% chance of belonging to the cluster. However, stars represented by faint open circles probably do not belong to M 7, since they have a less than 20% change of belonging to the cluster. Bold open circles may or may not belong to the cluster, since they have a membership probability between 20% and 90%. Reprinted from F. Gieseking (Astron. Astrophys. Suppl. Ser. 61:75–81, 1985) with permission.

NGC 6494 (M 23)

Constellation	Object type	RA, Dec	Approx. transit date at local midnight	Distance
Sagittarius	Open cluster	17h 57.1m, −18° 59′	July 15	2 thousand light years
Age	**Apparent size**	**Magnitude**	**Sky Atlas 2000.0 chart**	**Herald– Bobroff chart**
3 hundred million years	27′	5.5	15	C61/D11

Its size of 27′ corresponds to a diameter of about 15 light years. Professional telescopic studies count about 150 member stars that have less than a 10% chance of being "field stars" – see M 39.

NGC 6503

Constellation	Object type	RA, Dec	Approx. transit date at local midnight	Distance
Draco	Spiral galaxy	17h 49.5m, +70° 09′	July 13	17 million light years
Age	**Apparent size**	**Magnitude**	**Sky Atlas 2000.0 chart**	**Herald– Bobroff chart**
	7.0′ × 2.5′	10.2	3	C08

It is nearly edge-on (with an inclination angle of 74°). This is an unusually isolated galaxy, lying in the so-called "Local Void", a 15 million light year radius region containing few galaxies.

NGC 6514 (M 20)

Constellation	Object type	RA, Dec	Approx. transit date at local midnight	Distance
Sagittarius	Emission and reflection neb.	18h 02.7m, −22° 58′	July 16	5 thousand light years
Age	**Apparent size**	**Magnitude**	**Sky Atlas 2000.0 chart**	**Herald– Bobroff chart**
3 hundred thousand years	28′	6.3	22	C61/D11

Light from very young stars that formed out of the surrounding gas ionize hydrogen in the nebula causing it to glow (making the nebula a so-called HII region, although other elements are present, in particular oxygen, since the nebula benefits from a filter that lets in light from doubly ionized oxygen i.e. OIII). This is one of the youngest HII regions known, and is thought to be in a "pre-Orion-nebula" state, where young stars are violently ejecting matter, and protostars with jets are interacting with the nebula. The nebula is "lit up" mostly by the bright star in the middle of the nebula (HD 164492 or ADS 10991, mag. 7). ADS 10991 is a multiple star whose two brightest components have a separation of 10.6″ and position angle of 212°. It is found to consist of seven stars in professional telescopic studies. Dust grains are also present in the nebula (in the northern regions near the bright mag. 7.5 star there) which scatter the starlight, so that this nebula consists of both a reflection nebula (in the north) and an emission nebula (in the south). The nickname "Trifid Nebula" refers to the emission nebula, and is derived from Latin (trifidus, meaning "split into three"), which refers to its three lobes, which are separated by lanes of dust grains in the nebula that block its light from us. The entire nebula's size of 28′ corresponds to a diameter of about 40 light years.

NGC 6517

Constellation	Object type	RA, Dec	Approx. transit date at local midnight	Distance
Ophiuchus	Globular cluster	18h 01.8m, −08° 58′	July 16	70 thousand light years
Age	**Apparent size**	**Magnitude**	**Sky Atlas 2000.0 chart**	**Herald– Bobroff chart**
12–14 billion . years	1.0′	10.2	15	C44

It lies on the far side of our Galaxy, within the central bulge of our Galaxy (being about 10 thousand light years from the Galactic center). It is moderately metal-poor (i.e. it contains low amounts of elements heavier than helium), so that it may have formed in an early star-forming period of our Galaxy (see NGC 6287). Its size of 1.0′ gives it a diameter of about 20 light years. It has a mass of nearly 2 hundred thousand Suns.

NGC 6520

Constellation	Object type	RA, Dec	Approx. transit date at local midnight	Distance
Sagittarius	Open cluster	18h 03.4m, −27° 53′	July 16	5 thousand light years
Age	**Apparent size**	**Magnitude**	**Sky Atlas 2000.0 chart**	**Herald– Bobroff chart**
About 50 million years	6.0′	7.6	22	C61/D11

Its size of 6′ gives it a diameter of nearly 10 light years. The dark cloud to the west is Barnard 86 (i.e. B 86), or the "Inkspot Nebula", a dense cloud of dust grains between us and the rich background of old Milky Way stars.

NGC 6522

Constellation	Object type	RA, Dec	Approx. transit date at local midnight	Distance
Sagittarius	Globular cluster	18h 03.6m, −30° 02′	July 16	25 thousand light years
Age	**Apparent size**	**Magnitude**	**Sky Atlas 2000.0 chart**	**Herald– Bobroff chart**
12–14 billion years	5.6′	8.3	22	C61

It is one of the less massive globular clusters, with only 60 thousand solar masses. It lies just below (2 thousand light years below) the center of our Galaxy in Baade's window. This is a "window" near the center of our Galaxy where there is less obscuring dust, allowing us to see objects much closer to the center of our Galaxy in this window than in other regions. NGC 6522 is believed to have passed through the central Galactic plane about 2 million years ago. Its Galactic orbit, with a period of 10 million years or so, never takes it far from the Galactic center (always staying within about 5 thousand light years of it). It has a diameter of about 40 light years. This is a metal-poor cluster (meaning it has a low abundance of elements heavier than helium) and so may have formed in the earliest star-forming period of our Galaxy (see NGC 6287). It also has a collapsed core (see NGC 6284 for explanation).

NGC 6523 (M 8)

Constellation	Object type	RA, Dec	Approx. transit date at local midnight	Distance
Sagittarius	Emission nebula	18h 03.7m, −24° 23′	July 16	5 thousand light years
Age	**Apparent size**	**Magnitude**	**Sky Atlas 2000.0 chart**	**Herald–Bobroff chart**
	45′ × 30′	5.0	22	C61/D11

Nicknamed the "Lagoon Nebula". The open cluster NGC 6530 is embedded within this nebula. The stars in this cluster formed from the nebula in the past 2 million years or so. The nebula itself is a large ionized hydrogen (or HII) region, within which young O and B type stars associated with the cluster are thought to still be triggering star formation. M 8 is actually only a small "blister" on the surface of a giant molecular gas cloud that lies behind M 8. The nebula is ionized (and thus made visible) largely by just three stars (HD 165052, 9 Sgr, and HD 164740) as shown in Figure 6523. HD 164740 is a double star with a separation of 39″. It is responsible for ionizing the brightest part of the nebula, a 15″ × 30″ patch in the center of the nebula. This patch is called the "Hourglass" and is only about 10 thousand years old. The Hourglass is on the backside of the nebula, right on the edge of the giant molecular cloud from which NGC 6530 has formed.

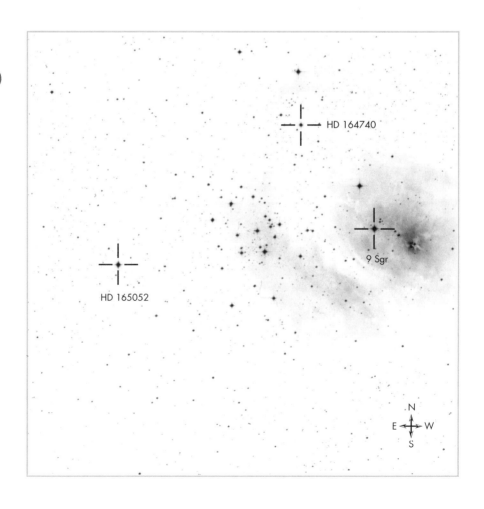

NGC 6528

Constellation	Object type	RA, Dec	Approx. transit date at local midnight	Distance
Sagittarius	Globular cluster	18h 04.8m, −30° 03′	July 18	30 thousand light years
Age	**Apparent size**	**Magnitude**	**Sky Atlas 2000.0 chart**	**Herald–Bobroff chart**
12–14 billion years	3.7′	9.6	22	C61

It has a diameter of about 30 light years. It lies close to NGC 6522, both apparently (being 16′ E of NGC 6522), and in actuality (lying 5 thousand light years away from NGC 6522), although it is nearly 60% more massive than NGC 6522 and contains much more "metal" (i.e. elements heavier than helium). Like NGC 6522, NGC 6528 lies in Baade's window (see NGC 6522 for explanation). It belongs to the central bulge of our Galaxy (whose properties are like those of elliptical galaxies) rather than the disk, and orbits the center of the Galaxy on a circular path at a distance of 10 thousand light years. It is "metal-rich" (meaning it has significant amounts of elements heavier than helium), so that it may have formed somewhat later than nearby metal-poor bulge clusters (see NGC 6287).

NGC 6531 (M 21)

Constellation	Object type	RA, Dec	Approx. transit date at local midnight	Distance
Sagittarius	Open cluster	18h 04.2m, −22° 29′	July 16	4 thousand light years
Age	**Apparent size**	**Magnitude**	**Sky Atlas 2000.0 chart**	**Herald–Bobroff chart**
10 milllion years	13′	5.9	22	C61/D11

This cluster lies very close to the Galactic central plane (being a few tens of light years below it). Of the 1.5 thousand or so open clusters in our Galaxy this is one of many that have not been very well studied.

NGC 6540

Constellation	Object type	RA, Dec	Approx. transit date at local midnight	Distance
Sagittarius	Globular cluster	18h 06.1m, −27° 46′	July 17	12 thousand light years
Age	**Apparent size**	**Magnitude**	**Sky Atlas 2000.0 chart**	**Herald–Bobroff chart**
12–14 billion years	0.8′	9.3	22	D11

It lies toward the center of our Galaxy in a smaller version of "Baade's window" (see NGC 6522 for explanation). Originally classified as an open cluster, it was reclassified as a globular cluster in the 1980s, making it the most recent addition to the list of NGC/IC globular clusters. It is a collapsed-core cluster (see NGC 6284 for explanation).

◄ **Figure 6523** (facing page) The region east of and including the emission nebula M 8 (NGC 6523) is shown (with the open cluster NGC 6530 in the middle of the figure), with the three stars responsible for ionizing M 8 marked with cross-hairs. The area shown is 30′ × 30′. From the Digitized Sky Survey (Space Telescope Science Institute) based on photographic data of the National Geographic Society – Palomar Observatory Sky Survey (POSS I).

NGC 6543

Constellation	Object type	RA, Dec	Approx. transit date at local midnight	Distance
Draco	Planetary nebula	17h 58.6m, +66° 38'	July 15	3 thousand light years
Age	**Apparent size**	**Magnitude**	**Sky Atlas 2000.0 chart**	**Herald–Bobroff chart**
	20"	8	3	C08

Nicknamed the "Cat's Eye Nebula". Its central star (mag. 10.9) is a Wolf–Rayet star (see NGC 40 for explanation). Professional telescopes find a very complex geometry to this nebula, consisting of a core (the bright region seen in amateur telescopes, about 20–25" in diameter) made of two expanding elliptical bubbles that are oriented perpendicular to each other, surrounded by nine concentric rings that are actually spherical shells thought to represent separate mass ejection events/shocks ejected over the past 5–20 thousand years at intervals of 1–2 thousand years. The edges of the core are pierced by fast-moving, low-ionization emission regions ("FLIERS" – see NGC 7662), placed symmetrically at the two ends of the major axis of the core. The inner nebula is surrounded by an outer halo (diameter 50') with an irregular outer edge. The sequence of events that created such an intricate structure to this nebula is not fully understood.

NGC 6544

Constellation	Object type	RA, Dec	Approx. transit date at local midnight	Distance
Sagittarius	Globular cluster	18h 07.3m, −25° 00'	July 17	8 thousand light years
Age	**Apparent size**	**Magnitude**	**Sky Atlas 2000.0 chart**	**Herald–Bobroff chart**
12–14 billion years	8.4'	7.8	22	C61/D11

This is one of the closest and most concentrated clusters known. It has a diameter of about 20 light years. It contains 130 thousand solar masses. It is one of a few (< 20) globular clusters that is known to contain a pulsar (within 5" of the optical center of the cluster), with the added oddity that the pulsar has a companion of planetary mass (10 times the mass of Jupiter, which is the second least massive known pulsar companion). The companion orbits the pulsar with a period of less than 2 hours, which is the second shortest period known for a binary pulsar. This is a collapsed-core cluster (see NGC 6284 for explanation).

NGC 6553

Constellation	Object type	RA, Dec	Approx. transit date at local midnight	Distance
Sagittarius	Globular cluster	18h 09.3m, −25° 54'	July 19	18 thousand light years
Age	**Apparent size**	**Magnitude**	**Sky Atlas 2000.0 chart**	**Herald–Bobroff chart**
12–14 billion years	3.2'	8.1	22	C61/D11

It has a mass of about 1/4 million Suns and a diameter of about 20 light years. It lies in the central bulge of our Galaxy, where it is patchily obscured by dust and gas that interferes with accurate study of this cluster. It has a circular orbit about the Galactic center with a radius of 10 thousand light years. It is "metal-rich" (meaning it has significant amounts of elements heavier than helium), so that it may have formed somewhat later than nearby metal-poor bulge clusters (see NGC 6287).

NGC 6568

Constellation	Object type	RA, Dec	Approx. transit date at local midnight	Distance
Sagittarius	Open cluster	18h 12.8m, –21° 35′	July 20	
Age	**Apparent size**	**Magnitude**	**Sky Atlas 2000.0 chart**	**Herald–Bobroff chart**
	13′	8.6	22	C61/D11

It was discovered in 1786 by William Herschel. Like most open clusters, it lies near the central plane of our Galaxy. Of the 1.5 thousand or so open clusters known in our Galaxy, distances are known for approximately 6 hundred. However, little information is known about this cluster, including its distance.

NGC 6569

Constellation	Object type	RA, Dec	Approx. transit date at local midnight	Distance
Sagittarius	Globular cluster	18h 13.6m, –31° 50′	July 19	30 thousand light years
Age	**Apparent size**	**Magnitude**	**Sky Atlas 2000.0 chart**	**Herald–Bobroff chart**
12–14 billion years	5.8′	8.6	22	C61

It lies in the central bulge of our Galaxy, just on the other side and slightly below (by several thousand light years) the center of our Galaxy. It has a mass of about 4 hundred thousand Suns and a diameter of nearly 50 light years.

NGC 6572

Constellation	Object type	RA, Dec	Approx. transit date at local midnight	Distance
Ophiuchus	Planetary nebula	18h 12.1m, +06° 51′	July 20	4 thousand light years
Age	**Apparent size**	**Magnitude**	**Sky Atlas 2000.0 chart**	**Herald–Bobroff chart**
	8″	8.8	15	C43

In professional telescopic studies, the elliptical shell of this nebula has disruptions that are thought to be due to more recent collimated mass ejections that are bipolar (i.e. the mass was ejected from the center of the nebula in opposite directions at the same time). Its size of 11″ corresponds to a diameter of about 0.2 light years.

NGC 6583

Constellation	Object type	RA, Dec	Approx. transit date at local midnight	Distance
Sagittarius	Open cluster	18h 15.8m, −22° 08′	July 21	
Age	**Apparent size**	**Magnitude**	**Sky Atlas 2000.0 chart**	**Herald–Bobroff chart**
	2.8′	10.0	22	C61/D11

It was discovered in 1786 by William Herschel. Little research has been done to date on this cluster.

NGC 6611 (M 16)

Constellation	Object type	RA, Dec	Approx. transit date at local midnight	Distance
Serpens Cauda	Open cluster with nebulosity	18h 18.8m, −13° 47′	July 20	6 thousand light years
Age	**Apparent size**	**Magnitude**	**Sky Atlas 2000.0 chart**	**Herald–Bobroff chart**
6 million years	7′	6.0	16	C61/D11

This open cluster is embedded in a gas cloud from which the cluster formed and in which star formation is still going on. Young, hot stars in the cluster are ionizing the surrounding hydrogen gas cloud (making it a so-called HII region), thereby making it fluoresce as the emission nebula IC 4703 which is nicknamed the "Eagle Nebula" or the "Star Queen Nebula" after the appearance of part of this nebula in professional telescopic images and photographs. This namesake region is well known from the Hubble Space Telescope's "Pillars of Creation" photo of a 2′ × 2′ or so portion of it (which lies just SE of the open cluster's most concentrated area). This photo shows three large pillars (looking like "elephant trunks" or hoodoos) aligned in a SE–NW direction. The pillars are regions of dark molecular gas and dust that are being "eroded" by intense radiation from stars to their NW. Professional telescopic studies have counted almost 4 hundred stars as members of the open cluster (but only a fraction of these are visible in amateur telescopes and are easily confused with field stars – see M 39). The cluster's size of 7′ corresponds to a diameter of about 10 light years.

NGC 6613 (M 18)

Constellation	Object type	RA, Dec	Approx. transit date at local midnight	Distance
Sagittarius	Open cluster	18h 19.9m, −17° 08′	July 21	5 thousand light years
Age	**Apparent size**	**Magnitude**	**Sky Atlas 2000.0 chart**	**Herald–Bobroff chart**
30 million years	9′	6.9	16	C61/D11

This sparse cluster contains about 20 members, has a diameter somewhat over 10 light years and has not been well studied.

NGC 6618 (M 17)

Constellation	Object type	RA, Dec	Approx. transit date at local midnight	Distance
Sagittarius	Emission nebula + open cluster	18h 20.8m, −16° 11′	July 21	7 thousand light years
Age	**Apparent size**	**Magnitude**	**Sky Atlas 2000.0 chart**	**Herald–Bobroff chart**
1 million years	11′	6.0	16	C61/D11

The appearance of this nebula in amateur telescopes leads to its various nicknames ("Swan Nebula", "Omega Nebula", among others). Its size of 11′ corresponds to a diameter of about 20 light years. Like M 8 and M 16, the nebula fluoresces due to an embedded open cluster (containing 35 or so stars) that formed from the nebula. However, the stars in the open cluster are so heavily obscured by intervening gas and dust that only five of them have magnitudes brighter than 14 (with only two brighter than magnitude 10), making its appearance as a true "cluster" of stars essentially nonexistent in the eyepiece of an amateur telescope.

NGC 6624

Constellation	Object type	RA, Dec	Approx. transit date at local midnight	Distance
Sagittarius	Globular cluster	18h 23.7m, −30° 22′	July 23	25 thousand light years
Age	**Apparent size**	**Magnitude**	**Sky Atlas 2000.0 chart**	**Herald–Bobroff chart**
12–14 billion years	5.9′	7.9	22	C61

It has a mass of a little over 3 hundred thousand Suns. It lies just below the center of our Galaxy in the central Galactic bulge. It is "metal-rich" (meaning it has significant amounts of elements heavier than helium), so that it may have formed somewhat later than nearby metal-poor bulge clusters (see NGC 6287). It has a diameter of over 40 light years. It contains an X-ray source caused by gas flowing from a white dwarf onto a neutron star that revolve around each other every 11 minutes at a separation of less than half the Earth–Moon distance! It is a collapsed-core cluster (see NGC 6284 for explanation).

NGC 6626 (M 28)

Constellation	Object type	RA, Dec	Approx. transit date at local midnight	Distance
Sagittarius	Globular cluster	18h 24.5m, −24° 52′	July 22	20 thousand light years
Age	**Apparent size**	**Magnitude**	**Sky Atlas 2000.0 chart**	**Herald–Bobroff chart**
	15′	6.8	22	C61/D11

Its mass is almost 1/2 million Suns. Like M 10 (NGC 6254), this cluster spends its time within a few thousand light years of the Galactic central plane and rotates approximately with the material in the Galactic disk. This is quite unusual for a "metal-poor" cluster like this one ("metal-poor" meaning it has low amounts of elements heavier than helium), since stars in the Galactic disk tend to have "metal" that was scattered by previous supernovae. The origin of this cluster is thus uncertain.

NGC 6629

Constellation	Object type	RA, Dec	Approx. transit date at local midnight	Distance
Sagittarius	Planetary nebula	18h 25.7m, –23° 12′	July 22	5 thousand light years
Age	**Apparent size**	**Magnitude**	**Sky Atlas 2000.0 chart**	**Herald–Bobroff chart**
	15″	11	22	C61/D11

It has a diameter of a few tenths of a light year. It is expanding outward at about 10 km/s. The central star is a mag. 12.9 Wolf–Rayet star (see NGC 40 for explanation of "Wolf–Rayet star").

NGC 6633

Constellation	Object type	RA, Dec	Approx. transit date at local midnight	Distance
Ophiuchus	Open cluster	18h 27.3m, +06° 31′	July 24	1 thousand light years
Age	**Apparent size**	**Magnitude**	**Sky Atlas 2000.0 chart**	**Herald–Bobroff chart**
6 hundred million years	4.6	27′	15	C43/D10

It was discovered in 1783 by Caroline Herschel (William Herschel's sister). Of the many hundreds of stars imaged in the cluster area by professional telescopes to date, about 70 stars have a > 50% probability of belonging to this cluster. It has a width of about 8 light years. It contains two known "blue stragglers", which are stars that are paradoxically far more blue and luminous than expected for reasons that remain unknown (but may involve the coalescence of stars). The two blue stragglers HD 170054 (mag. 8.2, 7′ SW of the cluster center) and HD 169959 (mag. 7.7, 15′ SW of the cluster center) are readily visible in amateur telescopes (see Figure 6633), or even binoculars.

NGC 6637 (M 69)

Constellation	Object type	RA, Dec	Approx. transit date at local midnight	Distance
Sagittarius	Globular cluster	18h 31.4m, –32° 21′	July 24	37 thousand light years
Age	**Apparent size**	**Magnitude**	**Sky Atlas 2000.0 chart**	**Herald–Bobroff chart**
12–14 billion years	7.1′	7.6	22	C61

It lies almost directly below the Galactic center (about 5 thousand light years below the Galactic central plane), about 10 thousand light years on the other side of the Galactic center from us. It has a mass of about 3 hundred thousand Suns and its size corresponds to a diameter of about 75 light years. It is thought to be a "bulge cluster" (meaning it spends its time orbiting around the central, spherical, 15 thousand light year diameter bulge of our Galaxy). It is a "metal-rich" cluster (meaning it contains significant amounts of elements heavier than helium). About 1/4 of the globular clusters in our Galaxy are considered metal-rich.

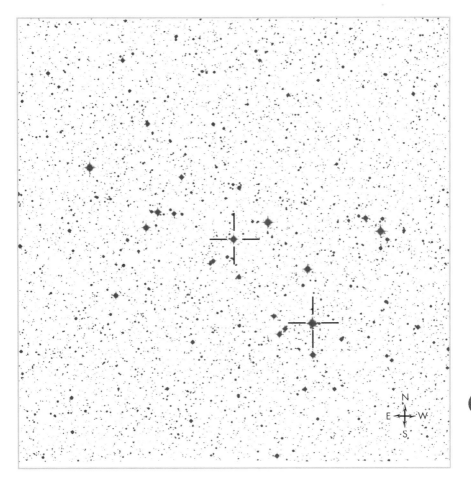

Figure 6633 The SW portion of the open cluster NGC 6633 is shown with the two blue stragglers HD 170054 (mag. 8.2, 7′ SW of the cluster center) and HD 169959 (mag. 7.7, 15′ SW of the cluster center) marked by the cross-hairs. The area shown is 30′ × 30′. From the Digitized Sky Survey (Space Telescope Science Institute) based on photographic data of the National Geographic Society – Palomar Observatory Sky Survey (POSS I).

NGC 6638

Constellation	Object type	RA, Dec	Approx. transit date at local midnight	Distance
Sagittarius	Globular cluste	18h 30.9m, −25° 30′	July 24	30 thousand light years

Age	Apparent size	Magnitude	Sky Atlas 2000.0 chart	Herald– Bobroff chart
12–14 billion years	2.2′	9.0	22	C61/D11

It is a medium-sized globular cluster of about 1 hundred thousand solar masses. Its size of 2.2′ gives it an actual diameter of about 20 light years.

NGC 6642

Constellation	Object type	RA, Dec	Approx. transit date at local midnight	Distance
Sagittarius	Globular cluster	18h 31.9m, –23° 29′	July 25	25 thousand light years
Age	**Apparent size**	**Magnitude**	**Sky Atlas 2000.0 chart**	**Herald–Bobroff chart**
12–14 billion years	0.8′	9.1	22	C61/D11

It lies close (in actual space) to NGC 6638, the two being 2–3 thousand light years apart, which is even closer together than NGC 6522 and 6528, although our line-of-sight makes the latter two appear closer together (16′ vs. 121′). Its size of 0.8′ gives it an actual diameter of 6 light years, making it one of the smaller size globulars. However, it has a medium mass for a globular (130 thousand Suns) so that it is a concentrated cluster. Like 3/4 of the globular clusters in our Galaxy, this is a metal-poor cluster (meaning it has a low abundance of elements heavier than helium).

NGC 6645

Constellation	Object type	RA, Dec	Approx. transit date at local midnight	Distance
Sagittarius	Open cluster	18h 32.6m, –16° 53′	July 24	
Age	**Apparent size**	**Magnitude**	**Sky Atlas 2000.0 chart**	**Herald–Bobroff chart**
	10′	8.5	15	C61/D11

It was discovered in 1786 by William Herschel. Little research has been done on this cluster to date.

NGC 6656 (M 22)

Constellation	Object type	RA, Dec	Approx. transit date at local midnight	Distance
Sagittarius	Globular cluster	18h 36.4m, –23° 54′	July 25	10 thousand light years
Age	**Apparent size**	**Magnitude**	**Sky Atlas 2000.0 chart**	**Herald–Bobroff chart**
12–14 billion years	17′	5.1	22	C61/D11

Its mass is about 1/2 million Suns. Along with M 15, this is one of only four globular clusters that contains a planetary nebula, labeled GJJC 1. However, at mag. 15 and lying near the core of the cluster, finding GJJC 1 in an amateur telescope is a challenge this author has not attempted (the one in M 15 is easier to find – see M 15). Recent professional telescopic observations of several stars in the bulge of our Galaxy on the other side of M 22 lying in the same field as M 22 from our line-of-sight ("field stars") showed transient brightening (by gravitational microlensing). One explanation for this is that a cluster of dark objects containing perhaps a million MACHOs (massive compact halo objects, like brown dwarves and planets) may lie between M 22 and the bulge of our Galaxy. However, it has also been suggested that planets within M 22 could be responsible for the microlensing – a definitive explanation awaits future research. M 22 never strays too far from the Galactic disk in its orbit, staying within about 5 thousand light years of the Galactic central plane between about 30 thousand and 10 thousand light years from the Galactic center, orbiting the Galaxy once every 2 hundred million years or so.

NGC 6664

Constellation	Object type	RA, Dec	Approx. transit date at local midnight	Distance
Scutum	Open cluster	18h 36.6m, –07° 49′	July 26	4 thousand light years
Age	**Apparent size**	**Magnitude**	**Sky Atlas 2000.0 chart**	**Herald–Bobroff chart**
14 million years	16′	7.8	16	C43/D10

It has a diameter of about 20 light years. It contains EV Sct (mag. 10), a spectroscopic binary that is one of only two known binary stars in our Galaxy that consists of two Cepheid variable stars (the other being CE Cas, which unlike EV Sct, can be split into its two component stars in amateur telescopes – see NGC 7790).

NGC 6681 (M 70)

Constellation	Object type	RA, Dec	Approx. transit date at local midnight	Distance
Sagittarius	Globular cluster	18h 43.2m, –32° 18′	July 27	30 thousand light years
Age	**Apparent size**	**Magnitude**	**Sky Atlas 2000.0 chart**	**Herald–Bobroff chart**
12–14 billion years	7.8′	7.9	22	C61

It has a mass of about 2 hundred thousand Suns and a diameter of about 70 light years. Like its neighbor M 69, this is thought to be a "bulge cluster" (meaning it spends its time orbiting around the bulge of our Galaxy – see M 69/NGC 6637). Its core is thought to have "collapsed", the result of an instability that causes the stars in its core to confine themselves to an unusually small region (see NGC 6284).

NGC 6694 (M 26)

Constellation	Object type	RA, Dec	Approx. transit date at local midnight	Distance
Scutum	Open cluster	18h 45.2m, –09° 24′	July 27	5 thousand light years
Age	**Apparent size**	**Magnitude**	**Sky Atlas 2000.0 chart**	**Herald–Bobroff chart**
90 million years	15′	8.0	16	C43/D10

It has a diameter of about 20 light years. It lies about 250 light years below the Galactic central plane, which happens to be the furthest extent our Sun travels from the Galactic central plane (although the Sun currently sits about 50 light years above the Galactic central plane).

NGC 6705 (M 11)

Constellation	Object type	RA, Dec	Approx. transit date at local midnight	Distance
Scutum	Open cluster	18h 51.1m, −06° 16′	July 29	6 thousand light years
Age	**Apparent size**	**Magnitude**	**Sky Atlas 2000.0 chart**	**Herald– Bobroff chart**
250 million years	14′	5.8	16	C43/D10

Nicknamed the "Wild Duck Cluster" after the V-shaped outline (pointed east) that some of its brighter members make. Its mass is several thousand Suns, with 5 hundred members brighter than mag. 14. Its size of 14′ corresponds to a diameter of about 25 light years. A person in the middle would see a night sky with several hundred first mag. stars, each separated by < 1 light year. This is nearly as dense as some globular clusters. A significant number of field stars are present as foreground/background stars (see M 39).

NGC 6712

Constellation	Object type	RA, Dec	Approx. transit date at local midnight	Distance
Scutum	Globular cluster	18h 53.1m, −08° 42′	July 29	25 thousand light years
Age	**Apparent size**	**Magnitude**	**Sky Atlas 2000.0 chart**	**Herald– Bobroff chart**
12–14 billion years	4.3′	8.1	16	C43/D10

This is one of about 150 globular clusters in our Galaxy. It passed through the Galactic plane a few million years ago. Numerous such previous interactions with the Galactic disk are thought to have stripped away many of its lighter stars, leaving them behind in the Milky Way's halo (which is the large spherical region around the flat, spiral central Galactic plane). Indeed, perhaps 99% of its original mass has been stripped away in this manner, so that it is now but a remnant of what was once one of the most massive globular clusters in our Galaxy. It once had a mass of perhaps 10–20 million Suns, but now contains only roughly 2 hundred thousand solar masses. It is one of only a dozen or so globular clusters in our Galaxy that is a bright X-ray source, thought to be caused by interacting binary stars (i.e. a neutron star and a low-mass companion).

NGC 6715 (M 54)

Constellation	Object type	RA, Dec	Approx. transit date at local midnight	Distance
Sagittarius	Globular cluster	18h 55.1m, −30° 29′	July 30	80 thousand light years
Age	**Apparent size**	**Magnitude**	**Sky Atlas 2000.0 chart**	**Herald– Bobroff chart**
12–14 billion years	9.1′	7.6	22	C61

This globular cluster is not part of our Galaxy, but instead belongs to a nearby satellite galaxy (that goes by the cumbersome name "Sagittarius Dwarf Elliptical Galaxy"). It has been suggested that M 54 may actually be its nucleus. This companion galaxy is in the process of being gravitationally disrupted by our Galaxy. Models predict that it, along with M 54, will collide with the disk of our Galaxy in several tens of millions of years, having had a past such collision about 2 hundred million years ago. M 54 is one of the most massive known globular clusters with a mass of about 1.5 million Suns (which is about 1/40 the mass of the galaxy it belongs to). Its size of 7.6′ corresponds to a diameter of about 2 hundred light years.

NGC 6720 (M 57)

Constellation	Object type	RA, Dec	Approx. transit date at local midnight	Distance
Lyra	Planetary nebula	18h 53.6m, +33° 02'	July 29	2 thousand light years
Age	Apparent size	Magnitude	Sky Atlas 2000.0 chart	Herald–Bobroff chart
	1.2'	9	8	C25

Nicknamed the "Ring Nebula". This planetary nebula is thought to have been created about a thousand years ago when an old star blew off its outer layers. In professional telescopes the nebula extends out to nearly 4' in diameter (with its outer halo 5 thousand times dimmer in surface brightness than the ring). Its three-dimensional structure remains inconclusively understood, but its main part may consist of an ellipsoidal shell (like the skin of an air-ship/dirigible as in the Hindenberg or the Good Year blimp) that we are looking at nearly end-on. This shell is thought to be more dense at its mid-section (half-way along the "dirigible") so that the ring we see is merely a torus of denser material at the mid-section of the ellipsoidal shell. The denseness of the ring is thought to be a relic of the preferential ejection of mass by the central star in its equatorial plane (in a "superwind" – see M 27). The shell is expanding outward at about 50 km/s. Invisible (UV) radiation from the hot central star ionizes the atoms in the shell, and electrons recombining with these ionized atoms cause optical photons to be emitted that make the nebula visible to us.

NGC 6755

Constellation	Object type	RA, Dec	Approx. transit date at local midnight	Distance
Aquila	Open cluster	19h 07.8m, +04° 16'	August 3	5 thousand light years
Age	Apparent size	Magnitude	Sky Atlas 2000.0 chart	Herald–Bobroff chart
60 million years	7.5	15'	16	C43/D10

The diameter of this cluster is about 20 light years. It was discovered in 1785 by William Herschel. The open cluster Czernik 39 lies about 7' NNW.

NGC 6756

Constellation	Object type	RA, Dec	Approx. transit date at local midnight	Distance
Aquila	Open cluster	19h 08.7m, +04° 42'	August 3	11 thousand light years
Age	Apparent size	Magnitude	Sky Atlas 2000.0 chart	Herald–Bobroff chart
130 million years	4.0'	10.6	16	C43/D10

It has a diameter of about 13 light years and was discovered in 1791 by William Herschel.

NGC 6779 (M 56)

Constellation	Object type	RA, Dec	Approx. transit date at local midnight	Distance
Lyra	Globular cluster	19h 16.6m, +30° 11'	August 4	32 thousand light years
Age	**Apparent size**	**Magnitude**	**Sky Atlas 2000.0 chart**	**Herald–Bobroff chart**
12–14 billion years	5.0'	8.3	8	C24/D09

It has a mass of about 2 hundred thousand Suns and a diameter of about 50 light years. Although its orbit is roughly circular and lies nearly in the Galactic disk (inclined by only about 15° to the central plane), like almost all globular clusters in our Galaxy, it does not orbit with the disk material at a constant radius from the Galactic center. Instead its path is thought to take it out as far as about 40 thousand light years away from the Galactic center and within a few thousand light years of the Galactic center, although it takes about the same amount of time to complete an orbit as our Sun (i.e. 1/4 billion years).

NGC 6781

Constellation	Object type	RA, Dec	Approx. transit date at local midnight	Distance
Aquila	Planetary nebula	19h 18.4m, +06° 32'	August 4	5 thousand light years
Age	**Apparent size**	**Magnitude**	**Sky Atlas 2000.0 chart**	**Herald–Bobroff chart**
	11.8	1.8'	16	C42/D10

It is about 2–3 light years in diameter and is probably a few tens of thousands of years old. It has an ellipsoidal shape in recent professional telescopic studies and is expanding at several tens of km/s.

NGC 6802

Constellation	Object type	RA, Dec	Approx. transit date at local midnight	Distance
Vulpecula	Open cluster	19h 30.6m, +20° 16'	August 9	4 thousand light years
Age	**Apparent size**	**Magnitude**	**Sky Atlas 2000.0 chart**	**Herald–Bobroff chart**
1 billion years	3.2'	8.8	8	C24/D09

This is a relatively old open star cluster. With a diameter of a little over 3 light years, it also is one of the smallest open clusters. Old age causes clusters like this one to disintegrate as gravitational forces tear stars away from it over time.

NGC 6809 (M 55)

Constellation	Object type	RA, Dec	Approx. transit date at local midnight	Distance
Sagittarius	Globular cluster	19h 40.0m, –30° 58'	August 10	20 thousand light years
Age	**Apparent size**	**Magnitude**	**Sky Atlas 2000.0 chart**	**Herald–Bobroff chart**
12–14 billion years	19.1'	6.3	22	C60

It has a mass of about 1/4 million Suns and a diameter of about a hundred light years. It takes a little over a hundred million years or so to complete an orbit about our Galaxy, always staying within about 20 thousand light years of the Galactic center but never swinging closer than about 5 thousand light years from the Galactic center, all in a path that is highly inclined to the disk of our Galaxy (by about 50–60°). It contains 74 known "blue stragglers" (see NGC 6633), which are stars that are paradoxically far more blue and luminous than expected for reasons that remain unknown (but may involve the coalescence of stars).

NGC 6818

Constellation	Object type	RA, Dec	Approx. transit date at local midnight	Distance
Sagittarius	Planetary nebula	19h 44.0m, –14° 09'	August 11	6 thousand light years
Age	**Apparent size**	**Magnitude**	**Sky Atlas 2000.0 chart**	**Herald–Bobroff chart**
	22"	10	16	C42

Nicknamed the "Little Gem". It has a diameter of a little more than 1/2 light year. It is expanding outward at about 50 km/sec. The central star has a mass of about 0.6 Suns, corresponding to a progenitor star of about one solar mass. The nebula is thought to be about 9 thousand years old.

NGC 6819

Constellation	Object type	RA, Dec	Approx. transit date at local midnight	Distance
Cygnus	Open cluster	19h 41.3m, +40° 11'	August 11	8 thousand light years
Age	**Apparent size**	**Magnitude**	**Sky Atlas 2000.0 chart**	**Herald–Bobroff chart**
2.5 billion years	5'	7.3	8	C24/D08

This is an old cluster and one of the richest known open clusters in our Galaxy, with at least 2.9 thousand member stars in professional telescopic studies. It has a diameter of about 10 light years and a mass of 2.6 thousand Suns.

NGC 6823

Constellation	Object type	RA, Dec	Approx. transit date at local midnight	Distance
Vulpecula	Open cluster	19h 43.2m, +23° 18′	August 12	6 thousand light years
Age	**Apparent size**	**Magnitude**	**Sky Atlas 2000.0 chart**	**Herald–Bobroff chart**
7 million years	12′	7.1	8	C24/D09

The remains of the nebulous gas from which this very young cluster formed is barely visible as an emission nebula that surrounds the cluster. The cluster has a diameter of a little more than 20 light years, while its surrounding nebulosity extends to a visible diameter of 70 light years. It contains about 40 stars in professional telescopic studies (but these are massive stars, with many being considerably more than 10 times as massive as the Sun).

NGC 6826

Constellation	Object type	RA, Dec	Approx. transit date at local midnight	Distance
Cygnus	Planetary nebula	19h 44.8m, +50° 32′	August 12	3 thousand light years
Age	**Apparent size**	**Magnitude**	**Sky Atlas 2000.0 chart**	**Herald–Bobroff chart**
	25″	9	3	C06

It has a diameter of < 1/2 light year and is thought to be several thousand years old. The central star has a mass of about 0.6 times that of our Sun and has a surface temperature of 45 thousand K. At mag. 10.7 this is one of the brightest central stars of all planetary nebulae visible to us. Nicknamed the "Blinking Planetary" because the nebula "blinks" (i.e. appears and disappears) when viewed by direct and averted vision in rapid succession.

NGC 6830

Constellation	Object type	RA, Dec	Approx. transit date at local midnight	Distance
Vulpecula	Open cluster	19h 51.0m, +23° 06′	August 12	5 thousand light years
Age	**Apparent size**	**Magnitude**	**Sky Atlas 2000.0 chart**	**Herald–Bobroff chart**
40 million years	12′	7.9	8	C24/D09

It has a diameter of nearly 20 light years and was discovered in 1784 by William Herschel.

NGC 6834

Constellation	Object type	RA, Dec	Approx. transit date at local midnight	Distance
Cygnus	Open cluster	19h 52.2m, +29° 24′	August 14	7 thousand light years
Age	**Apparent size**	**Magnitude**	**Sky Atlas 2000.0 chart**	**Herald–Bobroff chart**
80 million years	5′	7.8	8	C24/D09

It has a diameter of about 10 light years and was discovered in 1784 by William Herschel.

NGC 6838 (M 71)

Constellation	Object type	RA, Dec	Approx. transit date at local midnight	Distance
Sagitta	Globular cluster	19h 53.8m, +18° 47′	August 14	12 thousand light years
Age	**Apparent size**	**Magnitude**	**Sky Atlas 2000.0 chart**	**Herald–Bobroff chart**
12–14 billion years	6.1′	8.2	16	C24/D09

This is one of the closest globular clusters to us. With a mass of less than about 40 thousand Suns and a diameter of a little over 20 light years, this is a sparse globular cluster. It is one of only a few globular clusters whose orbit stays in the disk of our Galaxy (so it is said to be a "disk cluster"). Its orbit is highly elliptical (with a minor to major axis ratio of 0.2), and it takes about 160 million years to complete one orbit around our Galaxy. This cluster contains significant amounts of elements heavier than helium (and so is said to be "metal-rich" – only about 1/4 of the globular clusters in our Galaxy are metal-rich, the rest being metal-poor).

NGC 6853 (M 27)

Constellation	Object type	RA, Dec	Approx. transit date at local midnight	Distance
Vulpecula	Planetary nebula	19h 59.6m, +22° 43′	August 15	1 thousand light years
Age	**Apparent size**	**Magnitude**	**Sky Atlas 2000.0 chart**	**Herald–Bobroff chart**
	5.8′	7.3	8	C24/D09

Nicknamed the "Dumbbell Nebula", although a partially eaten apple might be a better description of its actual appearance in amateur telescopes. Its bipolar (bowtie-shaped) nature occurs in < 20% or so of all planetary nebulae. The shape of such planetary nebulae is thought to begin with an aging giant star giving off a large amount of gas in a "superwind" (traveling at 10 km/s, emitting 10^{-4} solar masses/year) that is concentrated near the star's equator. Once the core of the old star is eventually exposed, a hot, fast wind (1 thousand km/s, emitting 10^{-9} solar masses/year) slams into the previously emitted gas that is concentrated in the equatorial regions. This results in preferential expansion of this wind in the polar directions and the two familiar bipolar lobes. The lobes are ionized by short-wavelength, nonvisible radiation from the central star, and re-emit this radiation in visible wavelengths. The central star (mag. 13.8) lies at the narrowest part of the "bowtie" shape. The nebula is several thousand years old and still expanding (at several tens of km/s). In professional telescopes M 27 is found to have an elliptical shell surrounding it, as well as an internal elliptical shell, so that the structure of this nebula is quite complex.

NGC 6864 (M 75)

Constellation	Object type	RA, Dec	Approx. transit date at local midnight	Distance
Sagittarius	Globular cluster	20h 06.1m, −21° 55′	August 17	60 thousand light years
Age	**Apparent size**	**Magnitude**	**Sky Atlas 2000.0 chart**	**Herald– Bobroff chart**
12–14 billion years	4.6′	8.5	23	C60

It has a mass of about 1/2 million Suns and a diameter of about 80 light years. Like 3/4 of the globular clusters in our Galaxy, it is "metal-poor" (i.e. it has a low abundance of "metals", which in astrophysics means elements heavier than helium). It lies on the other side of the Galaxy from us, well below the Galactic central plane (by about 30 thousand light years).

NGC 6866

Constellation	Object type	RA, Dec	Approx. transit date at local midnight	Distance
Cygnus	Open cluster	20h 03.9m, +44° 10′	August 17	5 thousand light years
Age	**Apparent size**	**Magnitude**	**Sky Atlas 2000.0 chart**	**Herald– Bobroff chart**
4 hundred million years	7′	7.6	9	C24/D08

It has a diameter of about 10 light years and was discovered in 1790 by William Herschel.

NGC 6882

Constellation	Object type	RA, Dec	Approx. transit date at local midnight	Distance
Vulpecula	Open cluster	20h 12.0m, +26° 29′	August 18	2 thousand light years
Age	**Apparent size**	**Magnitude**	**Sky Atlas 2000.0 chart**	**Herald– Bobroff chart**
1.5 billion years	7′	5.7	9	C24/D09

This is a very old cluster. It has a diameter of about 4 light years. Its small size is due to its old age, most of its stars having been stripped away over the years.

NGC 6885

Constellation	Object type	RA, Dec	Approx. transit date at local midnight	Distance
Vulpecula	Open cluster	20h 12.0m, +26° 29′	August 18	2 thousand light years
Age	**Apparent size**	**Magnitude**	**Sky Atlas 2000.0 chart**	**Herald– Bobroff chart**
1.5 billion years	7′	5.7	9	C24/D09

This is a duplicate entry of NGC 6882.

NGC 6888

Constellation	Object type	RA, Dec	Approx. transit date at local midnight	Distance
Cygnus	Emission nebula	20h 12.1m, +38° 21′	August 18	5 thousand light years

Age	Apparent size	Magnitude	Sky Atlas 2000.0 chart	Herald–Bobroff chart
	20′ × 10′		9	C24/D08

Nicknamed the "Crescent Nebula". Its shape is due to a fast (2 thousand km/s) stellar wind from a nearby star (HD 192163, mag. 7.5) plowing previously emitted material from the star (when it was a red supergiant) into an ellipsoidal shell that is hitting a still earlier shell of ejected material. See Figure 6888 for the position of HD 192163. The collision of the two shells breaks them up into clumps. Ionization by UV radiation from the star HD 192163 makes the clumps visible to us. HD 192163 is a Wolf–Rayet star – a massive star about 30 times the mass of the Sun that is near the end of its life (which will end in a supernova) and which is losing matter at a prodigious rate (i.e. about an Earth-mass/year) via its fast wind (see NGC 2403 for further explanation of Wolf–Rayet stars).

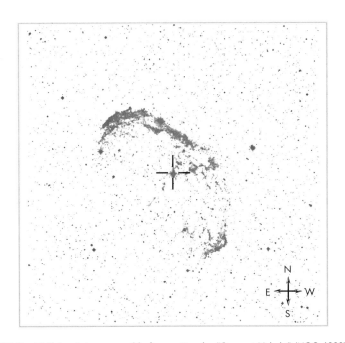

Figure 6888 The Wolf–Rayet star responsible for creating the "Crescent Nebula" (NGC 6888) is marked by the cross-hairs. The area shown is 30′ × 30′. From the Digitized Sky Survey (Space Telescope Science Institute) based on photographic data of the National Geographic Society – Palomar Observatory Sky Survey (POSS I).

NGC 6905

Constellation	Object type	RA, Dec	Approx. transit date at local midnight	Distance
Delphinus	Planetary nebula	20h 22.4m, +20° 06′	August 22	3 thousand light years
Age	**Apparent size**	**Magnitude**	**Sky Atlas 2000.0 chart**	**Herald–Bobroff chart**
	40″	12	16	C24/D09

It has a diameter of slightly more than 1/2 light year. Nicknamed the "Blue Flash" nebula. The central star (mag. 15.5) is a Wolf–Rayet star (see NGC 40 for explanation).

NGC 6910

Constellation	Object type	RA, Dec	Approx. transit date at local midnight	Distance
Cygnus	Open cluster	20h 23.2m, +40° 47′	August 22	4 thousand light years
Age	**Apparent size**	**Magnitude**	**Sky Atlas 2000.0 chart**	**Herald–Bobroff chart**
10 million years	8′	7.4	9	C24/D08

It has a diameter of a little under 10 light years. It is a very young open cluster.

NGC 6913 (M 29)

Constellation	Object type	RA, Dec	Approx. transit date at local midnight	Distance
Cygnus	Open cluster	20h 23.9m, +38° 32′	August 21	3 thousand light years
Age	**Apparent size**	**Magnitude**	**Sky Atlas 2000.0 chart**	**Herald–Bobroff chart**
A few million years	7′	6.6	9	C24/D08

It has a diameter of less than 10 light years. It is a very young cluster (young enough that circumstellar disks associated with planet formation could still be present – see NGC 2362). However, it is heavily obscured by foreground dust that is very patchy (dimming some stars in the cluster by up to 5 magnitudes, but hardly dimming others) making studies on it difficult.

NGC 6934

Constellation	Object type	RA, Dec	Approx. transit date at local midnight	Distance
Delphinus	Globular cluster	20h 34.2m, +07° 24'	August 25	50 thousand light years

Age	Apparent size	Magnitude	Sky Atlas 2000.0 chart	Herald–Bobroff chart
12–14 billion years	6.2'	8.8	16	C42

It lies far away and well below (20 thousand light years below) the Galactic central plane in the halo of our Galaxy. It orbits the Galaxy with a period of about a billion years, never coming any closer to the Galactic center than about 20 thousand light years, but wandering out about 150 thousand light years from the Galactic center at its most distant orbital position, on an orbit that is inclined to the Galactic disk by about 50°. This cluster moves in retrograde (i.e. opposite) to our Galaxy's rotation. It has an orbit that shares dynamical properties with that of NGC 5466 (see NGC 5466), and it has been suggested that both are left over from a disrupted satellite galaxy of the Milky Way. It is also moving toward us at high velocity (4 hundred km/s) compared to the surrounding objects in the halo. A size of 6.2' gives it an actual diameter of nearly 90 light years. It has a mass of about 2 hundred thousand Suns.

NGC 6939

Constellation	Object type	RA, Dec	Approx. transit date at local midnight	Distance
Cepheus	Open cluster	20h 31.5m, +60° 40'	August 24	9 thousand light years

Age	Apparent size	Magnitude	Sky Atlas 2000.0 chart	Herald–Bobroff chart
2 billion years	8'	7.8	3	C06

This is one of the older open clusters visible to amateur astronomers. It has a diameter of about 20 light years. The galaxy NGC 6946 is 38' SE (see NGC 6946). In professional telescopes it has more than 170 stars (i.e. stars with 750% probability of not being a field star).

NGC 6940

Constellation	Object type	RA, Dec	Approx. transit date at local midnight	Distance
Vulpecula	Open cluster	20h 34.4m, +28° 17'	August 25	2.5 thousand light years

Age	Apparent size	Magnitude	Sky Atlas 2000.0 chart	Herald–Bobroff chart
70 million years	31'	6.3	9	C24

It has a diameter of about 20 light years. It contains four known X-ray sources, all red giants, three of which are binary stars (which make up half of the six known binaries in this cluster).

NGC 6946

Constellation	Object type	RA, Dec	Approx. transit date at local midnight	Distance
Cygnus	Barred spiral galaxy	20h 34.8m, +60° 09'	August 25	17 million light years
Age	Apparent size	Magnitude	Sky Atlas 2000.0 chart	Herald–Bobroff chart
	8.8	11.2' × 9.8'	3	C06

It has an optical diameter of about 60 thousand light years. It rotates at about 2 hundred km/s. Recent star formation is present throughout the spiral arms. In fact, one of its spiral arms contains a gargantuan young star cluster (15 million years old) with more than a million solar masses, which is thought to be a young globular cluster. It is located in the knot 2.7' W and 2.0' S of the galaxy nucleus, and appears star-like in large amateur telescopes (see Figure 6946). This galaxy has the largest number of observed supernovae of any galaxy, with seven recorded supernovae. This galaxy lies in the so-called "Local Void", a 15 million light year radius region containing few galaxies.

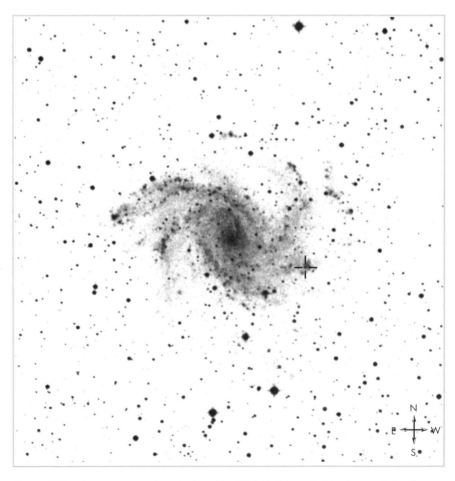

Figure 6946 A faint, super star cluster in the galaxy NGC 6946 is marked by the cross-hairs. The area shown is 15' × 15'. From the Digitized Sky Survey (Space Telescope Science Institute) based on photographic data of the National Geographic Society – Palomar Observatory Sky Survey (POSS I).

NGC 6960

Constellation	Object type	RA, Dec	Approx. transit date at local midnight	Distance
Cygnus	Supernova remnant	20h 45.7m, +30° 43′	August 27	1.4 thousand light years
Age	**Apparent size**	**Magnitude**	**Sky Atlas 2000.0 chart**	**Herald–Bobroff chart**
	70′ × 6′		9	C24/D08

This is the western portion of the two brightest segments of the "Veil Nebula" (the other, eastern, portion being NGC 6992 + NGC 6995), which are all part of the remains of a supernova that in professional telescopic studies is spread out over a 2.8° × 3.5° region of the sky (that also includes the central wisps NGC 6979 and NGC 6974). The nebula is the result of a supernova of a star with perhaps 10–15 times the mass of our Sun that occurred 5–10 thousand years ago. Prior to the supernova, the stellar winds from the star cleared out a cavity (or "bubble") in the interstellar material. The blast wave of the supernova is only now hitting the wall of this cavity where interstellar material lies. As the blast wave hits the interstellar material, it propagates as shock waves in the interstellar material (with the nonuniform, clumpy nature of the interstellar material causing these shock waves to have an intricate structure). As the shock waves pass through the interstellar material (at about 150 km/s), they heat material in the interstellar clouds to high temperatures. The subsequent cooling of the hot interstellar material (behind the shock wave) gives rise to radiation that we see as the nebula. Regions where a shock wave is propagating nearly parallel to the sky (i.e. viewed edge-on) are brighter (since then we are viewing a sheet of heated material edge-on). In contrast, a shock wave propagating nearly toward or away from us is dimmer (since then we are viewing a sheet of heated material face-on). The 70′ × 6′ apparent size of the nebula corresponds to dimensions of 30 × 2 light years.

NGC 6981 (M 72)

Constellation	Object type	RA, Dec	Approx. transit date at local midnight	Distance
Aquarius	Globular cluster	20h 53.5m, −12° 32′	August 30	65 thousand light years
Age	**Apparent size**	**Magnitude**	**Sky Atlas 2000.0 chart**	**Herald–Bobroff chart**
12–14 billion years	5.1′	9.3	16	C41

It has a mass of about 2 hundred thousand Suns and a diameter of about 1 hundred light years. It lies in the halo of our Galaxy (see M 2 for the meaning of "halo"). Like 3/4 of the globular clusters in our Galaxy, this is a "metal-poor" cluster (meaning it has a low abundance of elements heavier than helium). It rotates about our Galaxy in a retrograde direction (i.e. opposite to the Sun's motion around the Galaxy), which has led to the suggestion that it was adopted in a merger with another galaxy, but this is not certain.

NGC 6992/6995

Constellation	Object type	RA, Dec	Approx. transit date at local midnight	Distance
Cygnus	Supernova remnant	20h 56.4m, +31° 43′	August 30	1.4 thousand light years
Age	**Apparent size**	**Magnitude**	**Sky Atlas 2000.0 chart**	**Herald–Bobroff chart**
	72′ × 8′		9	C23

This is the eastern portion of the Veil Nebula (see NGC 6960).

NGC 6994 (M 73)

Constellation	Object type	RA, Dec	Approx. transit date at local midnight	Distance
Aquarius	Asterism	20h 58.9m, −12° 38′	August 31	
Age	**Apparent size**	**Magnitude**	**Sky Atlas 2000.0 chart**	**Herald–Bobroff chart**
	2.8′	9	16	C41

Although there has been some confusion as to whether this is an open cluster or not, recent data shows that the four stars at this location are not a cluster but are simply an asterism (i.e. a pattern of physically unrelated stars on the sky).

NGC 7000

Constellation	Object type	RA, Dec	Approx. transit date at local midnight	Distance
Cygnus	Emission nebula	21h 59.0m, +44° 31′	August 31	2 thousand light years
Age	**Apparent size**	**Magnitude**	**Sky Atlas 2000.0 chart**	**Herald–Bobroff chart**
	3° × 2°	4	9	C23/D07

Nicknamed the "North American Nebula" from its apparent shape in low-power amateur telescopes (with a nebula filter) or binoculars. If all its light were emitted by a point source, it would be appear as a magnitude 4 star in our sky. Of course, in actuality the light from this nebula is spread over several square degrees in the sky, so its brightness is quite dim (and nothing like a magnitude 4 star – see the Introduction for further explanation of the meaning of magnitude). NGC 7000 and the nearby "Pelican Nebula" (IC 5070, whose pelican shape is seen in photographic images) are both part of a single ionized hydrogen (HII) star-forming region (with a mass of about 5 thousand Suns) that is part of a single dark, giant molecular cloud (with a mass of about 50 thousand Suns). Part of the giant molecular cloud lies in front of the region between the two nebulae, making it appear as if the two nebulae are separate when in fact they are merely the outlying areas of an underlying emission region. It is the shape of the dark cloud superposed on the underlying (bright) HII region that produces the namesake appearances of these two nebulae. The underlying HII region is the part of the giant molecular cloud that has been ionized by a few hot young stars that formed within the cloud. The dim open cluster NGC 6997 that appears embedded in the nebula, approximately where Lake Superior would be, is actually a background object (about 2 hundred light years behind it) and did not evolve from the North American Nebula. This cluster is sometimes mistakenly labeled as NGC 6996.

NGC 7006

Constellation	Object type	RA, Dec	Approx. transit date at local midnight	Distance
Delphinus	Globular cluster	21h 01.5m, +16° 11′	September 1	140 thousand light years
Age	Apparent size	Magnitude	Sky Atlas 2000.0 chart	Herald–Bobroff chart
	2.2′	10.6	16	C23

It lies well out in the halo of our Galaxy (the halo is the region outside the spiral disk and bulge of our Galaxy, extending out as a sphere with a radius of perhaps six times that of the spiral disk region and containing most of the Galaxy's dark matter). Objects in the halo do not rotate with the spiral region, but instead have large, independent velocities. Indeed, NGC 7006 is moving at nearly 3 hundred km/s with respect to the Galactic rest frame and has one of the highest relative velocities of all globular clusters, approaching us at nearly 4 hundred km/s. It takes about 2 billion years to complete an orbit about the Galaxy, with its closest approach to the Galactic center being 60 thousand light years, and its most distant orbital position being over 3 hundred thousand light years from the Galactic center. It is currently moving toward the Galactic center. It has a mass of 250 thousand Suns. With a radius of 90 light years, this is a relatively large diameter globular, as is typical of globular clusters that are well out in the halo.

NGC 7008

Constellation	Object type	RA, Dec	Approx. transit date at local midnight	Distance
Cygnus	Planetary nebula	21h 00.6m, +54° 33′	August 31	3 thousand light years
Age	Apparent size	Magnitude	Sky Atlas 2000.0 chart	Herald–Bobroff chart
	1.4′	12	3	C05

It has a diameter of just over 1 light year. Professional telescopic studies show that the central star is a binary system. The nebula has an odd shape in the eyepiece, having been likened to a horseshoe with the open end to the southeast.

NGC 7009

Constellation	Object type	RA, Dec	Approx. transit date at local midnight	Distance
Aquarius	Planetary nebula	21h 04.2m, −11° 22′	August 31	3 thousand light years
Age	Apparent size	Magnitude	Sky Atlas 2000.0 chart	Herald–Bobroff chart
	29″	8	16	C41

Nicknamed the "Saturn Nebula" after the ansae that extend from the east and west edges in professional and large amateur telescopes. The two projections that give the nebula its name are FLIERs ("fast low-ionization emission regions") placed symmetrically on either side of the nebula. Although FLIERs occur in about half of all planetary nebulae, their origin remains poorly understood. In NGC 7009 they are thought to represent newly ejected material. The nebula is about 2 thousand years old and expanding at several tens of km/s. The magnitude 12.8 central star is visible in large amateur telescopes. In three dimensions and ignoring the ansae, the nebula has the shape of an elliptical shell. We are looking at the outside of this shell, so we see an oval shape. Its size of 0.5′ corresponds to a diameter of nearly 1/2 light year. In professional telescopes, this shell is contained within a giant halo (with a complex structure) that extends out to a diameter of well over 3′.

NGC 7027

Constellation	Object type	RA, Dec	Approx. transit date at local midnight	Distance
Cygnus	Planetary nebula	21h 07.0m, +42° 14'	September 2	3 thousand light years
Age	**Apparent size**	**Magnitude**	**Sky Atlas 2000.0 chart**	**Herald–Bobroff chart**
	15"	9	9	C23/D07

This nebula consists of an elliptical shell of ionized gas. In professional telescopic studies, this shell is found to be surrounded by an hourglass (bipolar) structure (perhaps due to two collimated jets) and an outer spherical halo of molecular gas that is in the process of being ionized by the central star, the latter indicating that this is a very young planetary (5 hundred to 1 thousand years old).

NGC 7044

Constellation	Object type	RA, Dec	Approx. transit date at local midnight	Distance
Cygnus	Open cluster	21h 13.2m, +42° 30'	September 3	10 thousand light years
Age	**Apparent size**	**Magnitude**	**Sky Atlas 2000.0 chart**	**Herald–Bobroff chart**
2 billion years	6'	12.0	9	C23/D07

This is one of the older NGC open clusters visible to amateur astronomers. It has a diameter of nearly 20 light years.

NGC 7062

Constellation	Object type	RA, Dec	Approx. transit date at local midnight	Distance
Cygnus	Open cluster	21h 23.5m, +46° 23'	September 6	5 thousand light years
Age	**Apparent size**	**Magnitude**	**Sky Atlas 2000.0 chart**	**Herald–Bobroff chart**
3 hundred million years	7'	8.3	9	C23/D07

It has a diameter of about 10 light years. This cluster has at least eight δ Scuti type pulsating variable stars, an uncommonly large number. These stars can be used like Cepheid variables (see NGC 7790) to give accurate estimates of distance based on their period, luminosity and color. The pulsations of δ Scuti variables also give rise to motions at the star surface of a few km/s, which can be used to probe the star's interior structure much like seismic waves on Earth give us information about the Earth's interior. It contains about 1 hundred stars in professional telescopic studies.

NGC 7078 (M 15)

Constellation	Object type	RA, Dec	Approx. transit date at local midnight	Distance
Pegasus	Globular cluster	21h 30.0m, +12° 10′	September 7	34 thousand light years
Age	**Apparent size**	**Magnitude**	**Sky Atlas 2000.0 chart**	**Herald–Bobroff chart**
12–14 billion years	12.3′	6.2	17	C41

Its mass is nearly a million Suns. Its size of 12.3′ corresponds to a diameter of 120 light years. It is a halo cluster (see M 2 for the meaning of "halo"), but never travels farther than about 45 thousand light years from the Galactic center on a path that is inclined by about 40° from the Galactic disk. It revolves once around the Galaxy every 1/4 billion years or so in a prograde orbit (like most globular clusters), meaning it revolves about the Galaxy in the same direction as the Galaxy's own rotation. The cluster is core collapsed (see NGC 6284 for explanation) and has one of the most concentrated centers (with more than 30 stars per square arcsecond in professional telescopes). It has been proposed that the cluster may contain a central black hole (with a mass of several thousand Suns), but this remains uncertain. M 15 does contain a planetary nebula (Pease 1, mag. 13), one of only four globular clusters that share this distinction and the easiest planetary of the four to find in amateur telescopes (but recommended for a 12-inch or larger telescope, and requiring a detailed map of the field, optional nebula filter, and patience to discern Pease 1 among the myriad stars near it).

NGC 7086

Constellation	Object type	RA, Dec	Approx. transit date at local midnight	Distance
Cygnus	Open cluster	21h 30.5m, +51° 36′	September 8	4 thousand light years
Age	**Apparent size**	**Magnitude**	**Sky Atlas 2000.0 chart**	**Herald–Bobroff chart**
140 million years	9′	8.4	3	C05

It has a diameter of about 10 light years and was discovered in 1788 by William Herschel.

NGC 7089 (M 2)

Constellation	Object type	RA, Dec	Approx. transit date at local midnight	Distance
Aquarius	Globular cluster	21h 33.5m, −00° 49′	September 8	40 thousand light years
Age	**Apparent size**	**Magnitude**	**Sky Atlas 2000.0 chart**	**Herald–Bobroff chart**
12–14 billion years	11.7′	6.5	17	C41

Its mass is about 9 hundred thousand Suns, but many of these stars are more massive than the Sun so that the total number of stars is about 150 thousand. Its size of 11.7′ gives it a diameter of about 130 light years. It lies in the halo of our Galaxy. The halo is the region outside the spiral disk and bulge of our Galaxy, extending out as a sphere with a radius of perhaps six times that of the spiral disk region and containing most of the Galaxy's dark matter. M 2 orbits the Galaxy independently of the Galactic disk on an inclined orbit that wanders out over a hundred thousand light years from the Galactic center and then approaches within a few tens of thousands of light years of the Galactic center, taking the better part of a billion years to complete one revolution around the Galaxy.

NGC 7092 (M 39)

Constellation	Object type	RA, Dec	Approx. transit date at local midnight	Distance
Cygnus	Open cluster	21h 31.7m, +48° 26′	September 7	1 thousand light years

Age	Apparent size	Magnitude	Sky Atlas 2000.0 chart	Herald–Bobroff chart
3 hundred million years	32′	4.6	9	C05/D07

It has a diameter of about 10 light years. This cluster lies in a rich field that is "contaminated" with field stars (i.e. stars that happen to lie along the same line-of-sight but which are foreground or background stars) from the Milky Way. This "contamination" worsens the fainter the stars that are being considered. For example, about 80–90% of the magnitude 8–10 stars are true cluster members, but only about 20% of the magnitude 11 stars in this cluster are actual cluster members (the rest being field stars), while fewer than 10% of the mag. 12 stars are true cluster members. Distinguishing field stars from true open cluster members requires professional telescopic studies, but the fact that one is seeing a mix of field stars and true cluster members should be borne in mind when viewing an open cluster through the eyepiece.

NGC 7099 (M 30)

Constellation	Object type	RA, Dec	Approx. transit date at local midnight	Distance
Capricornus	Globular cluster	21h 40.4m, −23° 11′	September 9	25 thousand light years

Age	Apparent size	Magnitude	Sky Atlas 2000.0 chart	Herald–Bobroff chart
12–14 billion years	9′	7.2	23	C59

It has a mass of about 3 hundred thousand Suns. Like most globular clusters, it orbits the Galaxy on a path that is inclined to the Galactic disk (by 50°), taking about 160 million years to complete one revolution around the Galaxy, never straying farther than about 25 thousand light years from the Galactic center, but never approaching closer than about 10 thousand light years to the Galactic center. The core of this cluster has "collapsed" (see NGC 6284), making its central region like a swarm of angry bees suddenly placed into a small container. Its size corresponds to a diameter of over a hundred light years. It contains the highest known globular cluster concentration of "blue straggler" stars. These are stars that are paradoxically far more blue and luminous than expected for reasons that remain unknown, but may involve the coalescence of stars (see NGC 6633), with almost 50 such stars known in this cluster.

NGC 7128

Constellation	Object type	RA, Dec	Approx. transit date at local midnight	Distance
Cygnus	Open cluster	21h 44.0m, +53° 43′	September 11	13 thousand light years

Age	Apparent size	Magnitude	Sky Atlas 2000.0 chart	Herald–Bobroff chart
20 million years	3.1′	9.7	3	C05

It has a diameter of about 12 light years. It lies near the inner edge of the next Galactic spiral arm outward from us (the Perseus arm).

NGC 7129

Constellation	Object type	RA, Dec	Approx. transit date at local midnight	Distance
Cepheus	Open cluster + reflection nebula	21h 43.0m, +66° 07'	September 10	3 thousand light years
Age	**Apparent size**	**Magnitude**	**Sky Atlas 2000.0 chart**	**Herald–Bobroff chart**
	2.7'	11.5	3	C05

This is a reflection nebula seen against a molecular cloud. It is a star-forming region, and several hot, young stars are visible as an embedded open cluster that has a very young age of about 1 hundred thousand years.

NGC 7142

Constellation	Object type	RA, Dec	Approx. transit date at local midnight	Distance
Cepheus	Open cluster	21h 45.2m, +65° 46'	September 12	5 thousand light years
Age	**Apparent size**	**Magnitude**	**Sky Atlas 2000.0 chart**	**Herald–Bobroff chart**
2 billion years	4.3'	9.3	3	C04

It has a diameter of about 6 light years and was discovered in 1794 by William Herschel. It is one of the older open clusters visible in amateur telescopes.

NGC 7160

Constellation	Object type	RA, Dec	Approx. transit date at local midnight	Distance
Cepheus	Open cluster	21h 53.7m, +62° 36'	September 14	3 thousand light years
Age	**Apparent size**	**Magnitude**	**Sky Atlas 2000.0 chart**	**Herald–Bobroff chart**
20 million years	7'	6.1	3	C04

It has a diameter of about 6 light years. Massive earlier generation stars in this cluster are thought to be responsible for a large shell of gas known as the Cepheus bubble, visible in professional telescopic studies as a 1 hundred light year diameter emission region that includes the nebulae Sharpless 2–131 (Sh 2–131), with embedded open cluster IC 1396, and Sharpless 2–140 (Sh 2–140).

NGC 7209

Constellation	Object type	RA, Dec	Approx. transit date at local midnight	Distance
Lacerta	Open cluster	22h 05.1m, +46° 29′	September 17	4 thousand light years

Age	Apparent size	Magnitude	Sky Atlas 2000.0 chart	Herald–Bobroff chart
4 hundred million years	25′	7.7	9	C04

It has a diameter of almost 30 light years. It contains the star SS Lac (mag. 10 – see Figure 7209), which was once a variable star of eclipsing Algol type. However, SS Lac is a triple star system. In such systems, the plane of the orbit of each star can change with time. Indeed, gradual changes in the orientation of the planes of the orbits of the three stars with respect to each other resulted in cessation of the eclipses in the middle of the twentieth century, so that this is no longer a variable star from our line-of-sight. This is only one of three eclipsing variable stars that is known to have stopped varying.

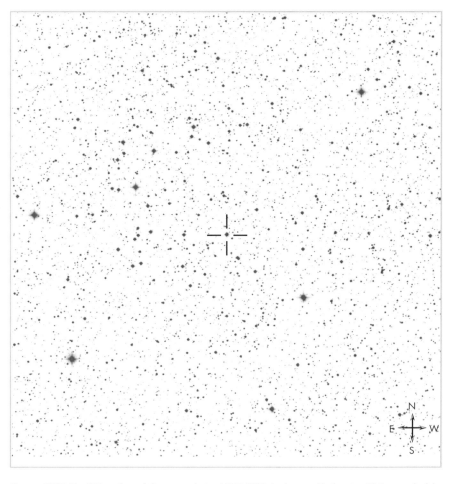

Figure 7209 The SW portion of the open cluster NGC 7209 is shown with the star SS Lac marked by cross-hairs. The area shown is 30′ × 30′. From the Digitized Sky Survey (Space Telescope Science Institute) based on photographic data of the National Geographic Society – Palomar Observatory Sky Survey (POSS I).

NGC 7217

Constellation	Object type	RA, Dec	Approx. transit date at local midnight	Distance
Pegasus	Spiral galaxy	22h 07.9m, +31° 22′	September 17	50 million light years
Age	**Apparent size**	**Magnitude**	**Sky Atlas 2000.0 chart**	**Herald–Bobroff chart**
	4.0′ × 3.4′	10.1	9	C22

This is an uncommon spiral in that it has an outer ring structure in professional telescopes, which is seen more commonly in galaxies with bars (which this galaxy does not have). In fact, three outer rings are found, located where density waves are in resonance with the local speed of epicycle oscillations in the orbits of matter in the disk (including the so-called Lindblad resonances). One explanation for the rings is that they were set up when a bar was present in the galaxy, but the bar has since disappeared, either due to a collision with another galaxy or by tidal interactions within the galaxy itself. Another strange feature of this galaxy is that 20–30% of its stars orbit in the opposite direction to the rest of the stars in the disk, a fact which may be related to the phenomenon that led to the ring structure.

NGC 7243

Constellation	Object type	RA, Dec	Approx. transit date at local midnight	Distance
Lacerta	Open cluster	22h 15.2m, +49° 54′	September 19	3 thousand light years
Age	**Apparent size**	**Magnitude**	**Sky Atlas 2000.0 chart**	**Herald–Bobroff chart**
1 hundred million years	21′	6.4	9	C04

Controversy exists as to whether this is even a cluster, since most stars previously considered by amateur astronomers as part of this cluster appear to be field stars. If considered a cluster, it contains an abnormally high number of chemically peculiar stars with strong magnetic fields, called CP2 stars (also referred to as Ap or Bp stars – see NGC 2169) which have abnormal line strengths in their spectra at certain wavelengths, thought to be associated with a separation of chemical elements in different layers at the surfaces of these stars.

NGC 7293

Constellation	Object type	RA, Dec	Approx. transit date at local midnight	Distance
Aquarius	Planetary nebula	22h 29.6m, −20° 50′	September 23	7 hundred light years

Age	Apparent size	Magnitude	Sky Atlas 2000.0 chart	Herald–Bobroff chart
	12.8′	7	23	C58

Nicknamed the "Helix Nebula" (from its appearance as a helical loop in professional telescopic images). The nebula is thought to have a mass of at least 1.5 Suns. In three dimensions the nebula is believed be disk shaped, about 1 light year thick and 3 light years in diameter, with the disk close to face-on (so we see it as round). The disk is much thicker (and brighter) at its outer edges, making it appear ring-like (particularly in professional telescopes). The mag. 13.5 central star is a hot white dwarf (with a temperature of about 120 thousand K, and a mass near that of our Sun) and is visible in amateur telescopes. The star that produced the nebula is thought to have had a mass of perhaps 6.5 times that of our Sun and would have had a visual magnitude of 5.5 in the sky, having lived about 60 million years before creating the nebula a little more than twenty thousand years ago (the nebula is thus an old one for a planetary nebula). The presence of an X-ray source suggests the central star may have a dwarf binary companion (with a mass of perhaps 20% that of our Sun and orbiting 3 times closer than Mercury is to our Sun), but this is not certain.

NGC 7296

Constellation	Object type	RA, Dec	Approx. transit date at local midnight	Distance
Lacerta	Open cluster	22h 28.1m, +52° 16′	September 23	

Age	Apparent size	Magnitude	Sky Atlas 2000.0 chart	Herald–Bobroff chart
	4′	9.7	3	C04

NGC 7295 is considered a duplicate entry of NGC 7296.

NGC 7331

Constellation	Object type	RA, Dec	Approx. transit date at local midnight	Distance
Pegasus	Spiral galaxy	22h 37.1m, +34° 25′	September 25	50 million light years

Age	Apparent size	Magnitude	Sky Atlas 2000.0 chart	Herald–Bobroff chart
	10.2′ × 4.2′	9.5	9	C22

It is about 150 thousand light years in diameter. It is thought to contain a supermassive black hole in its nucleus with a mass of a billion Suns. It is part of the so-called "Deer Lick" group of galaxies (named after Deer Lick Gap in the mountains of North Carolina). This is not an actual group of gravitationally bound galaxies, but rather simply several galaxies that happen to lie along the same line-of-sight, and which includes NGC 7335 (mag. 13.4), NGC 7336 (mag. 15), NGC 7337 (mag. 14.4) and NCG 7340 (mag. 13.7) that appear nearby (within 6′ in an easterly direction) but which are much farther away (3–4 hundred million light years) than NGC 7331.

NGC 7380

Constellation	Object type	RA, Dec	Approx. transit date at local midnight	Distance
Cepheus	Open cluster	22h 47.3m, +58° 08'	September 26	7 thousand light years

Age	Apparent size	Magnitude	Sky Atlas 2000.0 chart	Herald–Bobroff chart
10 million years	12'	7.2	3	C04

It has a diameter of about 25 light years. The remnants of the material from which this young cluster has formed can be seen as the faint nebulosity, labeled as Sharpless 2–142, in which this cluster is embedded. It was discovered in 1784 by Caroline Herschel (William Herschel's sister).

NGC 7448

Constellation	Object type	RA, Dec	Approx. transit date at local midnight	Distance
Pegasus	Spiral galaxy	23h 00.1m, +15° 59'	October 1	1 hundred million light years

Age	Apparent size	Magnitude	Sky Atlas 2000.0 chart	Herald–Bobroff chart
	2.6' × 1.2'	11.7	17	C40

It has an optical diameter of about 75 thousand light years and has more star-formation going on than in most spirals. It is the namesake member of the NGC 7448 group of galaxies, which probably contains nine galaxies, of which NGC 7454 (mag. 12), NGC 7463 (mag. 13) and NGC 7465 (mag. 13) can be readily seen within 1/2° to the east. The latter appears to have undergone a merger with another galaxy in the not too distant past, although at present no major interactions are occurring in this group.

NGC 7479

Constellation	Object type	RA, Dec	Approx. transit date at local midnight	Distance
Pegasus	Barred spiral galaxy	23h 04.9m, +12° 19'	October 2	110 million light years

Age	Apparent size	Magnitude	Sky Atlas 2000.0 chart	Herald–Bobroff chart
	4.0' × 3.1'	10.9	17	C40

It has a mass of a little over 2 hundred billion Suns and an optical diameter of about 130 thousand light years. The asymmetry in the spiral of this galaxy, and possibly even the bright bar itself, may be due to a recent minor merger with another galaxy in the recent past. Within the bar, which has a mass of approximately 50 billion Suns, vigorous star formation is believed to be occurring.

NGC 7510

Constellation	Object type	RA, Dec	Approx. transit date at local midnight	Distance
Cepheus	Open cluster	23h 11.1m, +60° 34'	October 4	9 thousand light years
Age	**Apparent size**	**Magnitude**	**Sky Atlas 2000.0 chart**	**Herald–Bobroff chart**
10 million years	4'	7.9	3	C04/D06

It has a diameter of about 10 light years and was discovered in 1787 by William Herschel. It lies on the edge of the Perseus arm of our Galaxy (which is the next spiral arm outward from us).

NGC 7606

Constellation	Object type	RA, Dec	Approx. transit date at local midnight	Distance
Aquarius	Spiral galaxy	23h 19.1m, –08° 29'	October 5	1 hundred million light years
Age	**Apparent size**	**Magnitude**	**Sky Atlas 2000.0 chart**	**Herald–Bobroff chart**
	4.2' × 2.3'	10.8	17	C40

It is one of fewer than about a hundred galaxies that have had two or more recorded supernovae.

NGC 7635

Constellation	Object type	RA, Dec	Approx. transit date at local midnight	Distance
Cassiopeia	Emission nebula	23h 20.8m, +61°13'	October 6	12 thousand light years
Age	**Apparent size**	**Magnitude**	**Sky Atlas 2000.0 chart**	**Herald–Bobroff chart**
	15' × 8'	9	3	C04/D06

Nicknamed the "Bubble Nebula" after its appearance in professional telescopic images. The "bubble" seen in professional telescopic images is about 3' in diameter (about 10 light years). It is caused by a fast stellar wind from the hot, young, central star (SAO 20575, mag. 8.7, lying near the center of the nebula as it appears in amateur telescopes) that is clearing out a roughly spherical cavity in the surrounding ionized hydrogen (HII) region. NGC 7635 is part of the much larger Sharpless 162 emission region (and which has a diameter of about 30'). Like other emission nebulae, such as M 8 and M 42, NGC 7635 is a small "blister" on a larger cloud of gas (with only the part that is ionized by the central star being visible). The amount of ionized gas in NGC 7635 is thought to be several solar masses.

NGC 7654 (M 52)

Constellation	Object type	RA, Dec	Approx. transit date at local midnight	Distance
Cassiopeia	Open cluster	23h 24.8m, +61° 36'	October 6	4 thousand light years
Age	**Apparent size**	**Magnitude**	**Sky Atlas 2000.0 chart**	**Herald–Bobroff chart**
1 hundred million years	13'	6.9	3	C03/D06

This is a relatively rich cluster. To magnitude 15.0, there is nearly one star for every square arcsecond of sky in its densest parts (although about 1 in 10 of these is a field star and not a cluster member – see M 39). To magnitude 14.5, a total of about 130 stars belong to the cluster (with only about 30 field stars "contaminating" the cluster), with these cluster members having masses about 2–5 times that of our Sun. To magnitude 19.5, over 6 thousand stars belong to the cluster (with about the same number of field stars present), with most of these dimmer cluster members having masses near that of our Sun. The stars in this cluster appear to have a much larger spread of ages (tens of millions of years) than most open clusters (where the stars are typically only a few million years apart in age). Gas and dust between us and the cluster dim the stars in this cluster considerably (by a few magnitudes).

NGC 7662

Constellation	Object type	RA, Dec	Approx. transit date at local midnight	Distance
Andromeda	Planetary nebula	23h 25.9m, +42° 32'	October 6	4 thousand light years
Age	**Apparent size**	**Magnitude**	**Sky Atlas 2000.0 chart**	**Herald–Bobroff chart**
	17"	9	9	C21

Nicknamed the "Blue Snowball". It has a double-ring structure (the inner ring brightened by a spherical shock wave, the outer due to ionization from the central star). It is a relatively small planetary, having an actual diameter less than a few tenths of a light year, corresponding to a size of 17". In professional telescopes, NGC 7662 appears to be one of about 50% of planetary nebulae that have a pair of fast-moving, low-ionization emission regions ("FLIERS") placed symmetrically on either side of the nebula (like small ears on a head). These regions are thought to be much younger than the remainder of the nebula and may be newly ejected material coming from deep inside the central star. The nebula is thought to be a little over a thousand years old.

NGC 7686

Constellation	Object type	RA, Dec	Approx. transit date at local midnight	Distance
Andromeda	Open cluster	23h 30.1m, +49° 08′	October 8	3 thousand light years
Age	**Apparent size**	**Magnitude**	**Sky Atlas 2000.0 chart**	**Herald–Bobroff chart**
	15′	5.6	9	C03

It has a diameter of about 13 light years and was discovered in 1787 by William Herschel.

NGC 7723

Constellation	Object type	RA, Dec	Approx. transit date at local midnight	Distance
Aquarius	Barred spiral galaxy	23h 39.0m, −12° 58′	October 10	80 million light years
Age	**Apparent size**	**Magnitude**	**Sky Atlas 2000.0 chart**	**Herald–Bobroff chart**
	3.5′ × 2.2′	11.2	17	C39

It has an optical diameter of about 80 thousand light years. Professional telescopes show the bar in this barred spiral has a radius of 23″ along the major axis of the galaxy. Stars at the ends of the bar are observed to be rotating at the same speed as a spiral wave there, so that the usual formation of new stars by density waves does not occur in these regions, causing star populations there to be older than at other locations. It is part of the NGC 7727 galaxy group (see NGC 7727).

NGC 7727

Constellation	Object type	RA, Dec	Approx. transit date at local midnight	Distance
Aquarius	Peculiar barred spiral galaxy	23h 39.9m, −12° 18′	October 11	80 million light years
Age	**Apparent size**	**Magnitude**	**Sky Atlas 2000.0 chart**	**Herald–Bobroff chart**
	4.7′ × 4.1′	10.6	17	C39

It is believed to be the result of two disk galaxies having merged approximately 1 billion years ago. The merger appears to have caused the formation of more than 20 young globular clusters within this galaxy. It forms a group of five gravitationally bound galaxies that includes nearby NGC 7723 (43′ SSW – see NGC 7723), along with two other dim non-NGC galaxies and the faint NGC 7724 (12′ WNW, mag. 13.5).

NGC 7789

Constellation	Object type	RA, Dec	Approx. transit date at local midnight	Distance
Cassiopeia	Open cluster	23h 57.4m, +56° 43′	October 15	8 thousand light years
Age	**Apparent size**	**Magnitude**	**Sky Atlas 2000.0 chart**	**Herald–Bobroff chart**
2 billion years	16′	6.7	3	C03/D06

Of the many thousands of stars imaged in the cluster area by professional telescopes, over 7 hundred stars have a > 50% probability of belonging to this cluster. It has a diameter of about 30 light years and was discovered in 1783 by Caroline Herschel (William Herschel's sister).

NGC 7790

Constellation	Object type	RA, Dec	Approx. transit date at local midnight	Distance
Cassiopeia	Open cluster	23h 58.4m, +61° 13′	October 15	11 thousand light years
Age	**Apparent size**	**Magnitude**	**Sky Atlas 2000.0 chart**	**Herald–Bobroff chart**
120 million years	17′	8.5	3	C03/D06

It has a diameter of about 50 light years. This is an unusual cluster because it contains an Algol type eclipsing binary (QX Cas, mag. 10.2) as well as three Cepheid variable stars: CF Cas (mag. 10.8–11.4, period 4.9 days), and CE Cas A & B (A: mag. 10.5–11.2, period 4.5 days, B: mag. 10.6–11.4, period 5.1 days). CE Cas is a binary system with separation 2.3″ and a position angle of about 80°. CE Cas is one of only two known double star systems in our Galaxy that consists of two Cepheids (the other is EV Sct in NGC 6664, but it is a spectroscopic binary). These Cepheids can be seen in amateur telescopes just west of the center of the cluster (see Figure 7790, overleaf), with the two stars making up CE Cas being discernable under good seeing conditions in amateur telescopes. Cepheids are variable stars whose luminosity varies in a regular manner due to periodic pulsations in the opacity of the star's atmosphere. These pulsations are the result of a cycle where radiative heating of the atmosphere causes expansion of the atmosphere, resulting in a more transparent atmosphere that leads to energy release, compression, and a more opaque atmosphere that starts the cycle again. The period is directly related to luminosity and so Cepheids are a standard means of determining astronomical distances.

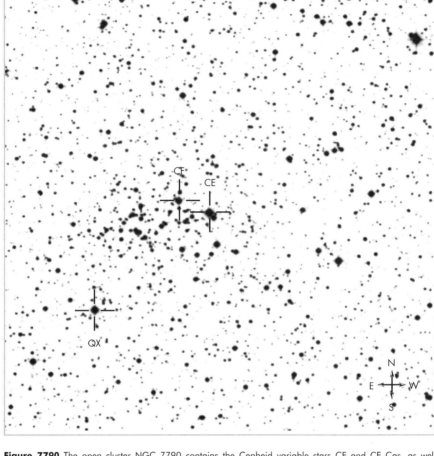

Figure 7790 The open cluster NGC 7790 contains the Cepheid variable stars CF and CE Cas, as well as the eclipsing binary star QX Cas, as marked. The area shown is 15′ × 15′. From the Digitized Sky Survey (Space Telescope Science Institute) based on photographic data of the National Geographic Society – Palomar Observatory Sky Survey (POSS I).

NGC 7814

Constellation	Object type	RA, Dec	Approx. transit date at local midnight	Distance
Pegasus	Spiral galaxy	00h 03.2m, +16° 09′	October 17	50 million light years
Age	**Apparent size**	**Magnitude**	**Sky Atlas 2000.0 chart**	**Herald–Bobroff chart**
	4.7′ × 2.4′	10.6	17	C21

This galaxy is almost exactly edge-on. Its size of 4.7′ gives it an actual diameter of about 70 thousand light years. Like the more well-known M 81, this galaxy has LINERs (low-ionization nuclear emission regions – see M 81/NGC 3031). Professional telescopes find its disk is warped (e.g. starting at 2.3′ from the center in the SE the dust lane bends to the SW in professional telescopic studies, twisting with the bend). Such warps are common, being observed in the neutral hydrogen (HI) distribution of about half of all galactic disks.

4

IC (Index Catalogue) Objects

IC 289

Constellation	Object type	RA, Dec	Approx. transit date at local midnight	Distance
Cassiopeia	Planetary nebula	03h 10.3m, +61° 19′	November 17	5 thousand light years
Age	**Apparent size**	**Magnitude**	**Sky Atlas 2000.0 chart**	**Herald– Bobroff chart**
	35″	12	1	C19

In professional telescopic studies this nebula has a ring-like structure inside two fainter spheroidal shells. The progenitor star is thought to have had a mass about twice that of our Sun and created the nebula within the past 10 thousand years. The nebula has a diameter of about a light year.

IC 4725 (M 25)

Constellation	Object type	RA, Dec	Approx. transit date at local midnight	Distance
Sagittarius	Open cluster	18h 31.8m, −19° 07′	July 24	2 thousand light years
Age	**Apparent size**	**Magnitude**	**Sky Atlas 2000.0 chart**	**Herald– Bobroff chart**
90 million years	29′	4.6	16	C61/D11

It has a diameter of about 20 light years and lies about 2 hundred light years below the Galactic central plane. It contains one Cepheid variable star (U Sgr, mag. 7, the right-hand star marked as U on Herald–Bobroff chart D11 – see NGC 7790 for explanation of Cepheid variables). It also contains six Be stars (see M 47 for explanation of Be stars).

Index

Note: Messier objects, IC objects and NGC objects with separate entries in Chapters 2–4 are not listed in the index since they are given in numerical order in the book and are therefore readily found.

AE Aurigae, 82
AGN (active galactic nucleus), 116
Andromeda galaxy, 20, 58
Antennae, the, 142
Ap star, 86, 87, 235

B 86, 205
Baade's window, 205, 207
Barnard 86, 205
Beehive Cluster, 25, 107
Be star, 26, 67, 69, 94, 101
bipolar planetary nebulae, 18, 221
Black-Eye Galaxy, 34, 178
blazar, 145
Blinking Planetary Nebula, 220
Blue Flash Nebula, 224
Blue Snowball, 239
blue straggler, 94, 95, 212, 213
Bubble Nebula, 238
bulge,
    Galactic, 4
    globular cluster, 36, 197, 212
Butterfly Cluster, 9, 201

Canes Venatici galaxy cloud, 145
Cat's Eye Nebula, 208
CE Cas, 241
Cepheid variable star, 18, 55, 77, 215, 241, 242, 243
CF Cas, 241
chemically peculiar star, 86, 87, 235
χ Persei, 69
Christmas Tree Cluster, 90
Clown Face Nebula, 98
collapsed-core globular cluster, 196
Cone Nebula, 90
contact binary stars, 186
CP2 star, 86, 87, 235
Crab Nebula, 7, 80
Crescent Nebula, 223
crossing time, 134
Czernik 20, 78
Czernik 39, 217

disk,
    circumstellar, 19, 97, 224
    counter-rotating galactic, 34, 128, 166, 178
    Galactic, 4
    globular cluster, 37, 221
Deer Lick group, 236
δ Scuti type star, 230
DL Cas, 55
Double Cluster, 68, 69
Duck Nebula, 95
Dumbbell Nebula, 18, 221

Eagle Nebula, 14, 210
early-type galaxy, 2, 52
Eridanus A galaxy group, 74
Eskimo Nebula, 98
ESO 495-G017, 107
E.T. Cluster, 62

field star contamination, 9–10, 23, 203, 232
Flame Nebula, 84
FLIERS (fast low-ionization emission regions), 239
flocculent spiral galaxy, 126

galaxy,
    counter-rotating disk, 34, 128, 166, 178
    early-type, 2, 52
    flocculent spiral, 126
    grand-design spiral, 28, 33, 48, 126, 132, 136, 139, 152, 182, 183
    inclination angle, definition, 5
    inner rings, 46, 121
    largest spiral, 80
    late-type spiral, 2, 52
    lenticular, 2
    Magellanic, 170
    multiple-armed spiral, 47, 139, 142, 149
    outer rings, 46, 111, 121, 159, 235
    supernovae rate, 41

galaxy (*continued*)
    with most number of globular clusters, 43
    with most number of recorded supernovae, 226
Ghost of Jupiter, 119
globular cluster,
    central black hole, 14, 231
    closest, 8, 37, 192, 208, 221
    collapsed-core, 196
    containing planetary nebulae, 14, 17, 214, 231
    fraction that are metal-poor, 37
    fraction that are metal-rich, 37
    highest concentration of blue stragglers in, 19, 232
    in bulge, 36, 197, 212
    in disk, 37, 221
    most concentrated, 14, 208, 231
    most luminous, 15, 100, 196
    most massive, 30, 202, 216
    nearest, 8, 37, 192, 208, 221
    outer halo, 195
    oldest, 45, 197, 198
    planetary nebula in, 14, 17, 214, 231
    retrograde orbit, 37
    shortest orbital period, 40, 191
    triple star system in, 8, 192
grand-design spiral galaxy, 28, 33, 48, 126, 132, 136, 139, 152, 182, 183
Gum 1, 93

HII (ionized hydrogen) region,
    largest known, 21, 64
    youngest, 16, 204
h Persei, 68
halo, Galactic, 4–5, 7, 229
Hercules Cluster, 13, 193
Helix Nebula, 236
Herbig-Haro object, 83
Holmberg IV, 48
Hourglass, the, 10, 206
Hubble's Variable Nebula, 90
Hyades, 21, 71, 107

ionized hydrogen (HII) region, largest known, 21, 64

youngest, 16, 204

IC objects mentioned but not given a separate entry,

IC 223, 69

IC 239, 70

IC 750, 143

IC 758, 142

IC 1029, 188

IC 1396, 233

IC 1590, 60

IC 2177, 93

IC 2574, 40

IC 3521, 159

IC 3583, 44

IC 4182, 180

IC 4213, 180

inclination angle, definition, 5

Inkspot Nebula, 205

inner rings, in galaxies, 46, 121

Intergalactic Wanderer, 100

interstellar spacing, 77, 105

interstellar extinction, 92, 101, 106, 111

Kemble's Cascade, 75

Lagoon Nebula, 10, 206

Large Magellanic Cloud, 170

late-type spiral galaxy, 2

lenticular galaxy, 2

Leo Triplet, 35, 130

Lindblad resonance, 46, 48, 126, 152, 153, 168, 176, 235

Little Ghost Nebula, 200

LINER (low ionization nuclear emission region), explanation of, 40, 114

Little Dumbbell Nebula, 38, 66

Little Gem, 219

Local Group, 20, 21, 53, 57, 58, 64

Local Void, 204, 226

Magellanic galaxy, 170

Markarian Chain, 42, 43, 155, 157, 158, 161

maser, 39, 72, 116

Mon R2 association, 86

multiple-armed spiral galaxy, 47, 139, 142, 149

μ Columbae, 82

NGC objects mentioned but not given a separate entry,

NGC 55, 59

NGC 300, 59

NGC 489, 63

NGC 502, 63

NGC 516, 63

NGC 518, 63

NGC 532, 63

NGC 586, 64

NGC 600, 64

NGC 636, 64

NGC 604, 64

NGC objects mentioned but not given a separate entry (continued)

NGC 660, 38

NGC 770, 68

NGC 899, 69

NGC 907, 69

NGC 941, 70

NGC 955, 70

NGC 988, 71

NGC 1003, 70

NGC 1023A, 70

NGC 1035, 71

NGC 1042, 71

NGC 1073, 39

NGC 1140, 71

NGC 1400, 74

NGC 1807, 78

NGC 2023, 39

NGC 2071, 39

NGC 2170, 86

NGC 2182, 86

NGC 2183, 86

NGC 2211, 88

NGC 2212, 88

NGC 2404, 99

NGC 2654, 109

NGC 2726, 109

NGC 2968, 112

NGC 2970, 112

NGC 3021, 114

NGC 3027, 113

NGC 3156, 117

NGC 3162, 118

NGC 3165, 117

NGC 3177, 118

NGC 3185, 118

NGC 3187, 118

NGC 3213, 118

NGC 3245A, 120

NGC 3254, 120

NGC 3265, 120

NGC 3287, 118

NGC 3299, 46

NGC 3301, 118

NGC 3338, 50

NGC 3346, 50

NGC 3381, 124

NGC 3389, 50

NGC 3396, 124

NGC 3413, 124

NGC 3424, 124

NGC 3625, 129

NGC 3430, 124

NGC 3442, 124

NGC 3507, 128

NGC 3592, 128

NGC 3599, 128

NGC 3605, 128

NGC 3611, 132

NGC 3630, 132

NGC 3641, 132

NGC 3642, 129

NGC 3648, 133

NGC 3652, 133

NGC 3658, 133

NGC 3659, 128

NGC 3664, 132

NGC 3669, 129

NGC 3674, 129

NGC 3681, 128

NGC objects mentioned but not given a separate entry (continued)

NGC 3683, 129

NGC 3684, 128

NGC 3691, 128

NGC 3718, 132, 134

NGC 3733, 136

NGC 3756, 136

NGC 3773, 135

NGC 3804, 136

NGC 3846A, 136

NGC 3850, 136

NGC 3896, 136

NGC 3899, 137

NGC 3906, 143

NGC 3913, 132

NGC 3972, 132

NGC 3990, 140

NGC 4020, 150

NGC 4027A, 141

NGC 4039, 142

NGC 4062, 150

NGC 4096, 143

NGC 4116, 146

NGC 4117, 143

NGC 4123, 146

NGC 4136, 150

NGC 4138, 143

NGC 4144, 51

NGC 4156, 145

NGC 4173, 150

NGC 4183, 143

NGC 4218, 143

NGC 4236, 40

NGC 4242, 51

NGC 4248, 51

NGC 4264, 150

NGC 4268, 150

NGC 4277, 150

NGC 4283, 150

NGC 4288, 143

NGC 4292, 33

NGC 4303A, 33

NGC 4310, 150

NGC 4359, 150

NGC 4389, 143

NGC 4395, 171

NGC 4437, 164

NGC 4458, 155

NGC 4460, 51

NGC 4461, 155

NGC 4469, 159

NGC 4476, 161

NGC 4525, 150

NGC 4562, 163

NGC 4568, 167

NGC 4624, 173

NGC 4625, 51, 170

NGC 4632, 174

NGC 4647, 172

NGC 4657, 173

NGC 4664, 173

NGC 4668, 174

NGC 4700, 175

NGC 4712, 175

NGC 4713, 172

NGC 4722, 175

NGC 4731, 174

NGC 4742, 175

NGC 4747, 175

Concise Catalog of Deep-sky Objects

NGC objects mentioned but not given a separate entry (*continued*)
NGC 4760, 177
NGC 4771, 172
NGC 4775, 174
NGC 4790, 175
NGC 4802, 175
NGC 4808, 172
NGC 4861, 180
NGC 4904, 172
NGC 4928, 179
NGC 4941, 174
NGC 4948, 174
NGC 4951, 174
NGC 4981, 179
NGC 5002, 180
NGC 5014, 180
NGC 5017, 181
NGC 5023, 34
NGC 5030, 180
NGC 5031, 181
NGC 5035, 181
NGC 5037, 181
NGC 5038, 181
NGC 5044, 181
NGC 5046, 181
NGC 5047, 181
NGC 5049, 181
NGC 5053, 29
NGC 5107, 180
NGC 5112, 180
NGC 5204, 48
NGC 5264, 41
NGC 5253, 41
NGC 5300, 185
NGC 5308, 184
NGC 5338, 185
NGC 5342, 184
NGC 5372, 184
NGC 5374, 185
NGC 5376, 184
NGC 5379, 184
NGC 5382, 185
NGC 5389, 184
NGC 5422, 186
NGC 5443, 186
NGC 5461, 48
NGC 5462, 48
NGC 5471, 48
NGC 5475, 186
NGC 5477, 48
NGC 5485, 186
NGC 5486, 186
NGC 5560, 187
NGC 5569, 187
NGC 5574, 187
NGC 5577, 187
NGC 5585, 48
NGC 5589, 187
NGC 5590, 187
NGC 5633, 188
NGC 5638, 189
NGC 5660, 188
NGC 5668, 189
NGC 5673, 188
NGC 5682, 188
NGC 5690, 189
NGC 5691, 189
NGC 5692, 189
NGC 5693, 188

NGC objects mentioned but not given a separate entry (*continued*)
NGC 5701, 189
NGC 5705, 189
NGC 5707, 188
NGC 5713, 189
NGC 5719, 189
NGC 5740, 189
NGC 5750, 189
NGC 5824, 195
NGC 5850, 189
NGC 5981, 191
NGC 5987, 191
NGC 5989, 191
NGC 6996, 228
NGC 6997, 228
NGC 7295, 236
NGC 7335, 236
NGC 7336, 236
NGC 7337, 236
NGC 7340, 236
NGC 7454, 237
NGC 7463, 237
NGC 7465, 237
NGC 7724, 240
NGC 7793, 59
North American Nebula, 228

oblique rotator, 86
Omega Nebula, 15, 211
open cluster,
    field star contamination, 9–10, 23, 203, 232
    nearest, 88
    number of in Milky Way, 209
    oldest, 35, 108
    richest, 219
    smallest, 218
Orion arm of Milky Way, 66
Orion A Complex, 24, 82
Orion B cloud, 39, 84
Orion-Monoceros Complex, 24, 82
Orion Nebula, 24, 82
outer rings, in galaxies, 46, 111, 121, 159, 235
Owl Cluster, 62
Owl Nebula, 47, 127

Pac-Man Nebula, 60
Perseus arm of Milky Way, 62, 66, 232
Pelican Nebula, 228
Pillars of Creation, 14, 210
Pinwheel Galaxy, 21, 47, 64, 149
planetary nebula, 1
    bipolar, 18, 221
    in globular clusters, 14, 17, 214, 231
    number of, in our Galaxy, 174
    quadrapolar, 103, 154
Pleiades, 25
Population I star, 4
Population II star, 5, 154
Praesepe, 25, 107
proplyds, 24, 83
pulsar, 7, 8, 33, 80, 192, 196, 199, 208

ram-pressure stripping, 45, 157, 162, 165
ρ Ophiuchi dust cloud, 192
Ring-Tail Galaxy, 142
Ring Nebula, 31, 217
Rosette Nebula, 89

Sagittarius Dwarf Elliptical Galaxy, 30, 188, 216
Saturn Nebula, 229
Seven Sisters, 25
Seyfert galaxy, 119
Sharpless 162, 238
Sharpless 2–131, 233
Sharpless 2–140, 233
Sharpless 2–142, 237
Siamese Twins, 167
Small Sagittarius Star Cloud, 17
Sombrero Galaxy, 50, 169
Spindle Galaxy, 49, 116, 190
SS Lac, 234
star,
    field, in open clusters, 9–10, 23, 203, 232
    Population I, 4
    Population II, 5, 154
    Wolf-Rayet, *see* Wolf-Rayet star
starburst galaxy, 41, 115
star formation, rate in Milky Way, 44, 168
Star Queen Nebula, 14, 210
Sun,
    distance from Galactic central plane, 13, 194
    location in Milky Way, 4
Sunflower Galaxy, 34, 168
super star cluster, 41, 60, 115, 183, 226
Swan Nebula, 15, 211
SZ Tau, 77

Tank Track Nebula, 84
τ CMa, 97
θ Aurigae, 86
Thor's Helmet, 95
Trapezium, 24, 82
triaxial, 178
Trifid Nebula, 16, 204

UGC galaxies,
    UGC 1195, 38
    UGC 1200, 38
    UGC 1519, 68
    UGC 1546, 68
    UGC 2275, 39
    UGC 2302, 39
    UGC 4549, 109
    UGC 5983, 125
    UGC 6791, 137
    UGC 6917, 141
    UGC 6922, 141
    UGC 6956, 141
    UGC 7000, 141
    UGC 7577, 159
    UGC 7698, 171
    UGC 8320, 34
    UGC 8837, 48

Ursa Major galaxy cluster, 52, 140

Veil Nebula, 227, 228
Virgo galaxy cluster, 27, 160
  brightest member in, 27, 160

largest and brightest spiral
  galaxy in, 48, 153
most disrupted galaxy in, 158

Whirlpool Galaxy, 28, 182

Wild Duck Cluster, 12, 216
Wolf-Rayet star,
  in NGC 2359, 95, 96
  in NGC 6888, 223
  in planetary nebulae, 55
Population I, 99